Energy Management System for Dispatchable Renewable Power Generation

Enhancing the integration of renewable power generation from wind and solar into the traditional power network requires mitigation of the vulnerabilities affecting the grid as a result of the intermittent nature of these resources. Variability and ramp events in power output are the key challenges to system operators due to their impact on system balancing, reserves management, scheduling, and the commitment of generation units.

This book presents the development of an energy management system for a renewable power generation (EMSRPG) tool that aims to achieve power-dispatching strategies based on forecasting renewable energy resource outputs to guarantee the optimal dispatch of hybrid wind-solar photovoltaic power systems (HWSPS). The key selling points of the book include the following:

- Renewable energy management in modern and future smart power systems
- Energy management systems
- Modeling and simulations using a real-time digital simulator (RTDS)
- High penetration level of renewable energy sources
- Case studies based on Oman's power systems and other power grids

This book discusses the challenges of integrating renewable resources, including low inertia systems, hosting capacity limitations of existing power systems, and weak grids. It further examines the detailed topologies, operation principles, recent developments in control techniques, and stability of power systems with a large scale of renewables. Finally, it presents case studies of recent projects from around the world where dispatchable power plant techniques are used to enhance power system operation.

Energy Management System for Dispatchable Renewable Power Generation

Edited by
Amer Al-Hinai and Hassan Haes Alhelou

CRC Press
Taylor & Francis Group
Boca Raton London New York

CRC Press is an imprint of the
Taylor & Francis Group, an **informa** business

First edition published 2023
by CRC Press
6000 Broken Sound Parkway NW, Suite 300, Boca Raton, FL 33487-2742

and by CRC Press
4 Park Square, Milton Park, Abingdon, Oxon, OX14 4RN

CRC Press is an imprint of Taylor & Francis Group, LLC

ISBN: 978-1-032-30958-3 (hbk)
ISBN: 978-1-032-30960-6 (pbk)
ISBN: 978-1-003-30743-3 (ebk)

DOI: 10.1201/9781003307433

Typeset in Times
by KnowledgeWorks Global Ltd.

Contents

Preface

The Sultanate of Oman has underutilized renewable energy (RE) due to its large reserves of oil and natural gas as these resources provide a cheaper source of power than renewable options. Population growth and the expansion of commercial and industrial sectors has increased electricity demand over the last 10 years by more than 240%. Average oil and gas production costs have increased over the last few years through the adoption of enhanced oil recovery and large-scale gas depletion techniques, and consumption of petroleum products has doubled in the last decade. In addition, the availability of gas for power generation is limited due to the decline of gas production and an increase in gas demand for other industries. These factors impose a challenge for the electricity sector in Oman and necessitate a need to diversify energy resources for power generation.

The Sultanate of Oman benefits from a long coastline and exposure to the strong summer and winter monsoon winds. It has an average wind speed of slightly more than 5 m/s and an estimated 2,463 hours of full load per year, making wind power an economically viable form of RE. One disadvantage of wind over solar is that it is more seasonal. Studies undertaken in Oman show that, in general, wind speed is higher during the summer months of June, July, and August and lower during October and November. As a result, wind speeds are higher during the months in which Oman reaches peak demand, which is a further indication of the feasibility of wind-powered electricity generation for the country. On the other hand, Oman has a high ratio of "sky clearness" and receives extensive daily solar radiation ranging from 5,500 to 6,000 Wh/m^2 a day in July to 2,500 to 3,000 Wh/m^2 a day in January, giving it one of the highest solar energy densities in the world. This puts the country in an ideal position to utilize both wind and solar as sustainable and alternative energy sources. Despite their potential, there are a number of challenges to utilizing RE resources in the Sultanate of Oman. These include high capital costs and output fluctuations. However, the recent development of RE technologies shows a declining trend in terms of cost and advancement in the integration of RE resources into existing power systems.

Enhancing the integration of renewable power generation from wind and solar into the traditional power network requires the mitigation of vulnerabilities that affect the grid as a result of the intermittent nature of these resources. Variability and ramp events in power output are key challenges to system operators due to their impact on system balancing, reserves management, scheduling, and the commitment of generation units. This book presents the development of an energy management system for a renewable power generation (EMSRPG) tool that aims to achieve power-dispatching strategies based on forecasting RE resource outputs to guarantee the optimal dispatch of hybrid wind-solar photovoltaic power systems (HWSPSs). The tool collectively utilizes forecasting models for wind speed and solar irradiance in addition to optimization techniques to size suitable wind and photovoltaic (PV) plant capacity. The energy storage system (ESS) is required to complement the HWSPS to achieve a robust and dispatchable plant. EMSRPG strategies aim to make full use of the complementary nature of wind and solar PV while using minimal ESS capacity to ensure power fluctuation mitigation and high power supply reliability. Based on profiles of wind speed and solar irradiance, the obtained results confirm that over 95% of available RE can be used.

Amer Al-Hinai
Hassan Haes Alhelou
Sultan Qaboos University

About the Authors

Dr. Amer Al-Hinai obtained his MSc and PhD degrees from West Virginia University. He is an associate professor in the Department of Electrical & Computer Engineering and the Deputy Vice-Chancellor for Postgraduate Studies & Research (DVC-PSR) at Sultan Qaboos University (SQU), Oman. He was a Visiting Professor at the Masdar Institute (MI) of Science and Technology, UAE, from 2012 to 2016. Dr. Al-Hinai was the founding Director of the Sustainable Energy Research Center (SERC) at SQU from 2017 to 2021. He has been a Senior Member of IEEE since 2011, Chairman of the IEEE–Oman Section from 2014 to 2018, and holds a Consultant grade as classified by the Oman Society of Engineers. He served as a Member and the Chairman of the Authority for Electricity Regulation, Oman, from 2011 to 2017. Dr. Al-Hinai has received many awards, including Recognition by SQU during the University Day (2013 and 2014) for Research and International award; Best Paper Award, UAE (2015); was one of the pioneers in the Engineering Practice in the Gulf, Bahrain (2011); and received the SQU Distinguished Researcher Award (2020). He has secured many internal and external research grants, notably from MI; The Research Council (TRC), Oman; His Majesty's Trust Fund (HMTF) for Strategic Research; British Petroleum (BP); Occidental Oman (OXY), and Petroleum Development Oman (PDO). His research interest spans diverse energy-related areas such as energy production, renewable energy resources, power systems, energy efficiency, and management and conversion. Dr. Al-Hinai has been a keynote and invited speaker on various aspects of renewable energy in Oman, UAE, and Switzerland. He is credited with more than 110 journals and conference papers. He is an associate editor of the *International Journal of Renewable Energy Technology*.

Dr. Hassan Haes Alhelou is a Senior Member of IEEE. He is affiliated with the Department of Electrical and Computer Systems Engineering, Monash University, Australia. He is also a professor and faculty member at Tishreen University in Syria and a consultant with Sultan Qaboos University (SQU) in Oman. Dr. Alhelou was with the School of Electrical and Electronic Engineering, University College Dublin (UCD), Ireland, from 2020 to 2021, and affiliated with Isfahan University of Technology (IUT), Iran. He completed his BSc from Tishreen University in 2011 and MSc and PhD from Isfahan University of Technology, Iran, all with honors. He was included in the 2018 and 2019 Publons and Web of Science (WoS), list of the top 1% best reviewers and researchers in the field of engineering and cross-fields all over the world. Dr. Alhelou was the recipient of the Outstanding Reviewer Award from many journals, e.g., *Energy Conversion and Management (ECM), ISA*

Transactions, and *Applied Energy*. He was the recipient of the best young researcher in the Arab Student Forum Creative among 61 researchers from 16 countries at Alexandria University, Egypt, in 2011. He also received the Excellent Paper Award 2021/2022 from the *IEEE CSEE Journal of Power and Energy Systems* (SCI IF: 3.938; Q1). Dr. Alhelou has published more than 200 research papers in high-quality peer-reviewed journals and international conferences. His research papers received 3500 citations with an h-index of 33 and an i-index of 56. He authored/edited 15 books published by reputed publishers such as Springer, IET, Wiley, Elsevier, and Taylor & Francis. He serves as an editor in a number of prestigious journals such as *IEEE Systems Journal, Computers and Electrical Engineering* (CAEE-Elsevier), *IET Journal of Engineering*, and Smart Cities. Dr. Alhelou has also performed more than 800 reviews for high-quality prestigious journals, such as *IEEE Transactions on Power Systems, IEEE Transactions on Smart Grid, IEEE Transactions on Industrial Informatics, IEEE Transactions on Industrial Electronics, Energy Conversion and Management, Applied Energy*, and the *International Journal of Electrical Power & Energy Systems*. He has participated in more than 15 international industrial projects across the globe. His major research interests are renewable energy systems, power systems, power system security, power system dynamics, power system cybersecurity, power system operation, control, dynamic state estimation, frequency control, smart grids, microgrids, demand response, and load shedding.

Contributors

Rashid Al-Abri
Department of Electrical and Computer
  Engineering
College of Engineering
Sultan Qaboos University
Muscat, Oman

Waleed Al-Abri
Department of Electrical and Computer
  Engineering
College of Engineering
Sultan Qaboos University
Muscat, Oman

Mohammed AlBadi
Department of Electrical and Computer
  Engineering
College of Engineering
Sultan Qaboos University
Muscat, Oman

Mohammed Al-Busaidi
Department of Electrical and Computer
  Engineering
College of Engineering
Sultan Qaboos University
Muscat, Oman

Hassan Haes Alhelou
Department of Electrical and Computer
  Engineering
College of Engineering
Sultan Qaboos University
Muscat, Oman

Muntaser Al Hasani
Department of Electrical and Computer
  Engineering
College of Engineering
Sultan Qaboos University
Muscat, Oman

Amer Al-Hinai
Department of Electrical and Computer
  Engineering
College of Engineering
Sultan Qaboos University
Muscat, Oman

Ahmed Al Maashri
Department of Electrical and Computer
  Engineering
College of Engineering
Sultan Qaboos University
Muscat, Oman

Mana Al-Shekili
Department of Electrical and Computer
  Engineering
College of Engineering
Sultan Qaboos University
Muscat, Oman

Abdullah Al Shereiqi
Department of Electrical and Computer
  Engineering
College of Engineering
Sultan Qaboos University
Muscat, Oman

Saira Al-Zadjali
Department of Electrical and Computer
  Engineering
College of Engineering
Sultan Qaboos University
Muscat, Oman

Mostafa Bakhtvar
Department of Electrical and Computer
  Engineering
College of Engineering
Sultan Qaboos University
Muscat, Oman

Myada Shadoul
Department of Electrical and Computer
    Engineering
College of Engineering
Sultan Qaboos University
Muscat, Oman

Edward Baleke Ssekulima
Department of Electrical Engineering
    and Computer Science
Masdar Institute
Abu Dhabi, UAE

Hassan Yousef
Department of Electrical and Computer
    Engineering
College of Engineering
Sultan Qaboos University
Muscat, Oman

Introduction

In most countries around the world, demand for electricity is growing rapidly. One of the challenges the electricity sector faces is meeting this demand by supplying customers with reliable and stable power. As a result, conventional power resources are being supplemented by renewable resources. Most power suppliers depend on fossil fuels as the primary energy source due to their ready availability and lower cost compared to other resources. However, an increase in demand, accompanied by heightened oil and gas production costs, is driving the use of other energy resources, such as renewable energy (RE). According to an International Renewable Energy Agency (IRENA) report, the vast majority of investment in RE is in wind and solar resources. In fact, the report states that in 2015, these two resources accounted for around 90% of all RE investment.

The main challenges of utilizing renewable resources are high capital costs and fluctuations in wind and solar power output. However, recent developments in RE technologies demonstrate a declining trend in cost. There have also been advancements in the integration of renewable resources into existing conventional power resources. This requires the mitigation of vulnerabilities imposed on the grid through the intermittent nature of these resources. Variability and ramp events in power output are key challenges for system operators due to their impact on system balancing, reserves management, scheduling, and the commitment of generation units. Forecasting accuracy and power output are highly correlated with each other. Thus, further efforts are needed to mitigate the risk associated with integrating renewable resources into the electricity grid.

Following the integration of renewable resources, the electricity grid must remain dispatchable and economically feasible. The prediction of renewable resources is a complex subject that has been investigated by numerous researchers. Forecasting models for mitigating the variability and unpredictability of solar and wind power resources are useful techniques with performance criteria depending on several different factors. The primary purpose of predicting renewable generation is defining the power output of generation plants as accurately as possible. Hence, studies focusing on the operation of wind farms and solar plants rely on predicted wind and solar profiles.

In this book based on industry and research projects, wind speed and solar irradiance prediction models are developed to aid in the design of hybrid wind-solar photovoltaic power systems (HWSPS) and the development of appropriate dispatching strategies. The seasonal autoregressive integrated moving average model coupled with a support vector machine can be used to develop the solar irradiance-forecasting model. A nonlinear autoregressive exogenous neural network model can be utilized to obtain wind speed forecasts using temperature as the exogenous input. Many countries have limits on allowable power fluctuations for grid-connected wind farms. For example, the Danish grid code allows for a maximum of 5% of the installed capacity per minute, while the UK National Grid Code allows for a maximum of 10 MW per minute. Furthermore, power plant operators must have their scheduled dispatch ready and bids submitted at least 1 hour ahead of the start of the half-hour trading period ("Gate Closure"), as is the case in the UK energy market.

While the dispatch of conventional power generation depends on an energy management system (EMS), such a tool has not yet been developed for RE generation. The EMS is a system of computer-aided tools used by load dispatch center operators to monitor, control, and optimize the performance of the generation and/or transmission system. Similarly, the EMS has been developed at the distribution system level to optimize the operation of distribution networks. Likewise, the EMS is established at the load side to enhance the efficiency of the connected load and energy storage system.

Renewable energy systems (RESs) have been used to displace energy from conventional generating plants but not to displace their capacity, as they are not visible to system operators. Therefore, available RE resources for power generation are used only as a means of reducing fossil-fuel consumption. This is mainly related to the inherently variable and nondispatchable nature of RE resources. Such a practice poses a threat to power system reliability and requires utilities to maintain power-balancing reserves to match the variable supply from RE resources and demand power levels. Maintaining these reserves for renewable generation represents an additional cost for the utility and jeopardizes the economic value of RE projects. On the other hand, escalating global energy demand and the increasing penetration of RESs have necessitated the development of strategies that can enable these intermittent resources to be dispatched reliably and economically. Only then can these resources truly displace fuel-based conventional generation in the electricity market and become the driving force of the future grid. Such problems should be solved with a properly designed EMS. The existing EMS focuses on the operation and control of conventional power generation and lacks the ability to accommodate the planning, operation, control, optimization, and dispatch of RE generation.

This book is prepared based on the output from His Majesty's Trust Fund (HMTF) research grant (Project code: SR/ENG/ECED/17/01) aimed at developing a centralized integrated renewable energy management system (EMSRPG) tool for planning, coordinating, and 100% reliably dispatching renewable energy resources for power generation. The project envisions a framework for novel technologies of EMSRPG; this book presents its novel findings in the field of energy management systems.

1 Literature Review on Renewable Energy Management Systems and Frequency Control

Muntaser Al Hasani, Amer Al-Hinai, Hassan Haes Alhelou, Hassan Yousef, Ahmed Al Maashri

CONTENTS

DOI: 10.1201/9781003307433-1

1.1 MICROGRID AND THE CHALLENGE OF FREQUENCY CONTROL

With the ever-increasing demand for electrical power across various industry types coupled with the rising focus on climate and environmental issues, power systems are witnessing a paradigm shift as they evolve from centralized power generation to a system that relies on a growing number of small power stations that rely on renewable energy sources (RESs).

In recent decades, the use of solar photovoltaic (PV) and wind resources has opened opportunities to exploit renewable resources in electric power production. In 2010, the RESs installed capacity over the globe reached 1,226.9 GW, and in 2019 this capacity increased by more than 207% to reach around 2,536.9 GW [1].

However, because of this inherent variability in output power of RESs, which instantaneously impacts the system frequency and continuity of demand and supply of power, the sophistication of power systems is growing. The rising share of RESs, which inherently have a significant amount of power electronics, thereby decoupling the power source from the grid as well as national scale decisions for retiring conventional power plants with synchronous machines, has serious effects on power system inertia, which is primarily responsible for stability and reliability of the power system.

In this context, a microgrid (MG) usually can be explained as a local area network of electrical power sources, such as renewable and conventional, as well as local loads, which can be operated in both MG operation modes to bridge the gap between the high increase in consumption and storage of energy and transmission issues. Given the fact that it is a scaled-down version of bigger electrical grids equipped with a self-sustaining energy system, it is typically employed for smaller geographic locations such as schools, colleges, or a local town.

Figure 1.1 illustrates the simple structure of MG. This hybrid system of wind turbines, solar PV, BESS, and DG, along with adequate management, will reduce energy production fluctuations and allow for grid interconnection.

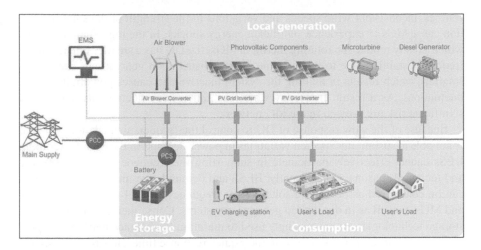

FIGURE 1.1   Concept overview of an MG [2]. EMS, energy management system; EV, electric vehicle, PCC, point of common coupling; PCS, power conversion system.

In an MG, aggregating the wind turbine output with solar PV generation requires incorporating the BESS and DG to stabilize the power imbalance caused by an RES output power variation. The battery energy storage system (BESS) will store additional energy and compensate for short and long power generation variability. MGs operate 24/7/365 handling and feeding energy to their customers. Hence, the three characteristics of an MG are as follows:

Local: It generates energy for consumers and overcomes the inefficiency of power losses (as in centrally located plants exporting power over long distances) by producing power locally to those it serves.

Independent: It can disconnect itself from the central grid in circumstances of cascading central grid network failure by islanding and still continue to supply its customers.

Intelligent: This comes from the MG controller responsible for managing the electrical power sources like solar PVs, wind doubly-fed induction generators (DFIGs), DGs, and BESS in real time based on defined conditions [3].

Although MGs have numerous advantages for power systems, they also present numerous obstacles, particularly in the areas of safety and control systems. The most significant problem in the use of RESs, such as wind and solar PV, is the fluctuation in active power produced as a result of variations in wind speed and sun irradiation variability. Consequently, avoiding frequency excursions in an island-based MG is the most significant issue in the area of MG studies.

According to [4], controlling MGs in the island operation mode is more complex than in the grid-connected mode because in the network mode, the system voltage and frequency settings of MGs are first maintained by the main grid network. Additionally, in an island MG that is powered by RESs such as wind and solar energy sources, inconsistent wind speeds and variations in the strength of the solar irradiation cause MGs to experience significant energy shocks. In the high share of RESs, which greatly reduce the rotating inertia in MGs, the DG has a smaller impact on the frequency stability due to its very low inertia (distribution generation's capacity is smaller), which results in making the system inactive and causes potentially severe fluctuations. As a result, frequency variations and voltage fluctuations that pose a significant risk are noticed in the MG network.

Because of the slow response of DG to control the frequency in islanded MGs, it is not able to sustain unexpected variations in power demand in the MG. In addition, RESs cannot effectively participate in supporting the frequency due to economic and technical reasons. Consequently, BESS and demand-side management systems can be considered as promising solutions for frequency control in modern systems and MGs, as well as in distributed generation systems. Consequently, battery energy storage technologies are increasingly used to maintain the balance between the production and consumption sides due to their quick response times, low time constants, and low ramp up/ramp down rates.

According to [4], "A number of important problems exist for MGs, including low inertia, uncertainty, complex dynamics, nonlinear structure, and intermittent RES. This means that when the load and power generation are out of balance, MGs will have a frequency and voltage variation (off from their normal values). MG blackouts are also possible when there is a mismatch between the load and the power generation." An effective controller strategy will help to ensure the system's balance between generation and demand.

1.2   WIND POWER GENERATION

Wind energy is produced by absorbing the mechanical energy of the wind and converting it to electrical energy using the fundamental technique of Faraday's electromagnetic theory. When it comes to renewable energy, wind power is one of the fastest-growing technologies on the globe. Since wind turbines and power converters continue to develop, the cost of generating electricity from wind has reduced dramatically.

Global wind power capacity reached 564 GW by the end of 2018. In 2019, the worldwide weighted-average cost of energy for new onshore wind farms was $0.053/kWh, with country/region values ranging from $0.051 to 0.099/kWh, according to [5]. This shows that the installation cost becomes reasonable when compared with the costs in the last decades.

1.2.1   DOUBLY-FED INDUCTION GENERATOR (DFIG)

Wind turbine units are constantly increasing in terms of both technology and installed capacity. Wind turbine technologies are classified according to their speed, and fixed-speed wind turbines and variable-speed wind turbines are the most common. Wind turbines are typically large, heavy, and slow-moving equipment. There are two types of wind turbine generators (WTGs) based on the type of their rotor structure, i.e., wound rotor winding and squirrel cage induction. These

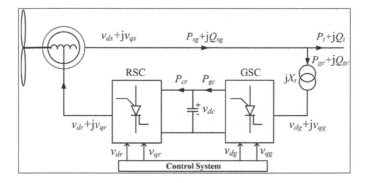

FIGURE 1.2   MG of a DFIG [6]. GSC, grid side converter; RSC, rotor side converter.

systems make use of a multistage gearbox that is coupled to a fixed-speed squirrel-cage induction generator (SCIG).

Mechanical energy is converted to electrical energy via the use of windings, and this is the basis for classifying electric machines. A single winding is used in a single-feed machine. These machines have a single winding that aids in the energy conversion process. Windings in double-fed machines contribute to energy conversion as well as the machine's overall efficiency. The wound-rotor DFIG is the only electric machine capable of running at twice the required torque at a given operating frequency [6].

There are direct connections between the stator's windings and the grid. Instead, an alternating current (AC)/direct current (DC) power converter is used to power the grid and a DC/AC converter is used on the rotor side to power the rotor windings. Parallel connections are made between the two AC/DC converters. In spite of the fact that these converters function together, they may not all have the same power output. The components of a DFIG are shown in Figure 1.2.

In essence, the wind turbine model is divided into three parts, i.e., the engine, the drive systems, and the converter units, when designing the generators is critical for the operation and use in modern power systems. The transformer, which has two secondary windings, one for each stator and rotor, is used to match the generator voltage to the grid side voltage.

DFIGs have the advantage of maintaining a consistent terminal voltage and frequency regardless of wind speed on the turbine rotor when utilized in wind turbines. As a result, they can operate over a wide variety of wind speeds while maintaining the appropriate tip-speed ratio, which ensures maximum efficiency. It is thus possible for the DFIG system to function at a range of rotor speeds that are both sub- and over-synchronous.

1.2.2   DFIG Power Converters

Figure 1.3 illustrates in detail the design of a DFIG-based wind turbine. In operation, the back-to-back converter behaves as a bidirectional power converter connected to a shared DC bus.

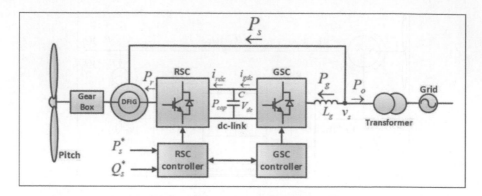

FIGURE 1.3   Active power flow in a DFIG turbine [7]. GSC, grid side converter; RSC, rotor side converter.

The DFIG's torque and speed, as well as the power factor at the stator terminals, are controlled by the rotor side converter (RSC) of the generator. The grid side converter (GSC) is responsible for maintaining a constant voltage in the DC connection, independent of the rotor power amplitude and rotational direction. As a result, the GSC's frequency is synced with the grid frequency (i.e., generating or absorbing reactive power). Back-to-back converters convert the fluctuating voltage and frequency generated by DFIG speed fluctuations into a constant voltage and frequency, allowing for grid connection at the point of common coupling (PCC), allowing for grid connectivity [8].

1.2.3   ROTOR SIDE CONVERTER (RSC)

Multiple control strategies can be utilized to manage the RSC [9]. However, because the stator is directly connected to the grid, the stator voltage-oriented control (SVOC) is employed because the frequency and voltage are assumed constant during normal operation. It is possible to construct the SVOC by coordinating the d-axis of the synchronous reference frame with the reference voltage of the stator:

$$V_{qs} = 0, \quad V_{ds} = V_s \tag{1.1}$$

In the same manner that the position of the rotor angle, θ_r, is referred to the stator reference frame, the position of the stator angle, θ_s, is referenced to the stator reference frame. It is the slip angle specified by Equation 1.1 that results from the angle formed between the voltage vectors of the stator and the rotor:

$$\theta_{sl} = \theta_s - \theta_r \tag{1.2}$$

Equation 1.2 allows for reciprocal frame transformations of abc and dq. As demonstrated in [10], the maximum power point tracking (MPPT) block control generates the electromagnetic torque, which controls the RSC via the current i_{dr}. Similarly, reactive power provides pulses to regulate the RSC via current i_{qr}.

1.2.4 GRID SIDE CONVERTER (GSC)

The GSC is accountable for maintaining a consistent voltage and frequency on the grid side. The GSC consumes or generates reactive power from the grid to control the voltage. GSC also uses two control loops to achieve its goal. The outer control loop controls DC voltage. The outer loop's output is a d-axis reference current $i_{cd,ref}$ that feeds the current regulating the inner loop. Inner and outer control loops use proportional-integral (PI) controllers to achieve zero steady-state error. The magnitude and phase of the GSC voltage are controlled by the voltage component of the GSC [11].

The corresponding circuit of DFIG is depicted in Figure 1.4.

The rotor current and quadrature components can be used to change the circuit's active and reactive power. The DFIG stator converter regulates the active and reactive power flow, which is the magnetizing current. The GSC concepts are seen in block diagrams that use current loops to i_d and i_q having i_d^* as a reference from the DC-link. Because $i_t^* = 0$, the convertor work at a power factor equals one.

Figure 1.5 is a detailed concept sketch of a wind turbine device. An electrical power converter is required, after which the torque of the generator is produced to extract the maximum power applied.

1.3 DIESEL-BASED POWER GENERATION

Diesel generators are often used in off-grid applications such as rural areas or onboard marine vessels for generating electricity or for backup power during electricity blackouts. Additionally, diesel generators are employed to regulate voltage and frequency.

Although diesel generators are not environmentally friendly, they are currently used in the majority of MG deployments because they are needed for MG frequency management because they provide the necessary inertia in the off-grid mode by operating in the isochronous mode.

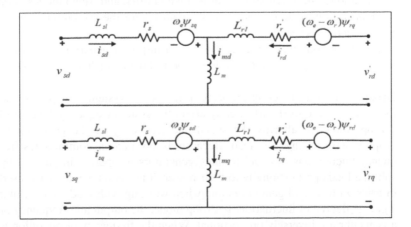

FIGURE 1.4 Block diagram of the equivalent circuit of DFIG [7].

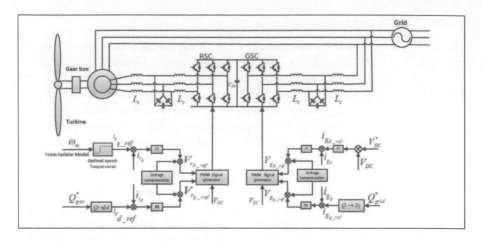

FIGURE 1.5   Control representations for a DFIG turbine [7].

According to [12], Rural Areas Electricity Company (RAECO; Tanweer), which is the only vertically integrated company in Oman, owns and manages power stations that are strategically located near load centers and settlements throughout the Sultanate of Oman. The total installed capacity is around 83 MW as of 2018.

In addition, RAECO (Tanweer) has initiated the development of 11 solar-diesel storage hybrid power projects in Oman totaling a capacity of 146 MW. According to plans, the 11 sites will have a total of 48 MW solar PV components and 70 MW of diesel components. A total of 28 MW of battery storage capacity is planned for the facilities, which will produce 14 MWh of storage capacity for electrical energy.

1.3.1   OPERATION MODES OF DIESEL ENGINE GENERATOR

From an application point of view, power generators can serve in the electrical power system installations using standalone, parallel with utility, and island modes of operation. Each operation mode imposes specific controls for the excitation system and governs the generator's speed.

The control methods that are commonly used in power generation sources like DGs for managing the power system voltage and frequency include the following:

  i. **Droop:** This is an inherent property of the power system. It explains that the change in the load will directly affect the system frequency. The word "droop" refers to altering the speed and frequency of the generator to proportionally change the load. Because droop mode accommodates frequency fluctuations, it enables many generators to operate in parallel by splitting loads proportionally to their power. It is advantageous when used in grids with several generators and when working with loads that exhibit a greater degree of fluctuation. In droop mode, the output and frequency of a generator are inversely proportional. When the frequency of operation is reduced, output increases. If a generator is set to 5% droop, for example, a

5% reduction in frequency increases the unit's output power by 100%. In this case, the frequency is increased by 1%, and the unit's power output is reduced by 20%.

ii. **Isochronous:** In this mode, the generator runs at its rated synchronous speed independent of the load demand. Because the isochronous mode lacks a power feedback link, it is usually only applicable when just one generator is running. A load feedback system is a necessity for managing a load of more than two generators when running parallel under this mode.

The isochronous method provides open-loop control to restore frequency deviations to their nominal value, whereas the droop mode provides closed-loop control to resolve transients and ensure the system is restored to its set point.

In MGs, the aim of the DG is to ensure the following:

1. In grid-connected mode, to interchange real and reactive power with the grid, the DG must operate in droop control mode, while the grid regulates voltage and frequency.
2. In off-grid (island) mode, the DG operates in isochronous operation mode to regulate the MG frequency by maintaining a constant speed.

1.3.2 EXCITATION AND SPEED CONTROLS OF DIESEL ENGINE GENERATOR

The primary controls of the DGs are their excitation and speed controls. The excitation system manages the synchronous generator's voltage and reactive power, whereas the speed control system regulates the speed (frequency) and active power of the prime mover and generators.

A block schematic of a typical DG system demonstrating two closed-loop response control systems for excitation and engine speed is shown in Figure 1.6.

Figure 1.7 shows the speed response control system and Figure 1.8 shows the excitation response control system. The combined representation of the two control systems is presented in Figure 1.9.

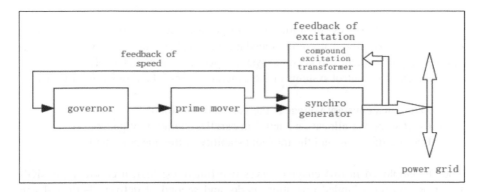

FIGURE 1.6   Block scheme of DG system [13].

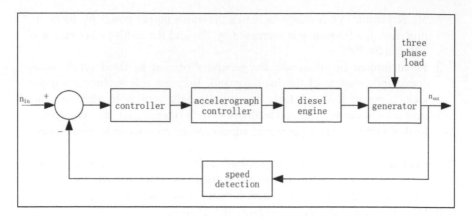

FIGURE 1.7 Diesel engine speed feedback control system [13].

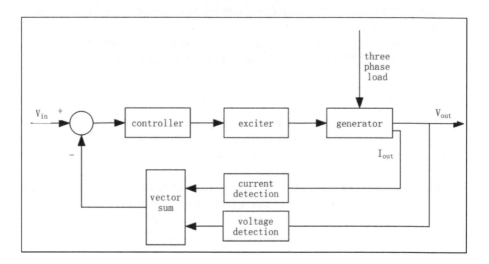

FIGURE 1.8 Generator excitation feedback control system [13].

The speed control system of the DG manages the frequency and the active power output of the generator. The engine speed must be kept constant to maintain the generator's output frequency to keep the system operated stably and safely.

The excitation control systems of the generators serve the following purposes:

    i. Maintain a constant voltage at the generator terminals
   ii. Reactive power allocation control in parallel operation of generators
  iii. Improve the static and the transient stability of the power system

The DG employed in this chapter serves two functions: first, it controls the MG's frequency during islanded operation mode, and second, it delivers both real and reactive power during on-grid and off-grid operation modes of the MG.

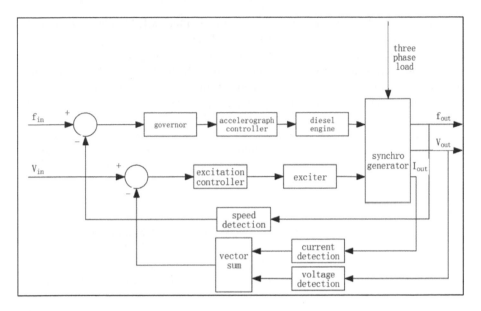

FIGURE 1.9   Frequency and voltage feedback control system of DG [13].

In the isochronous mode, the DG supplies a significant amount of inertia to sustain the MG nominal frequency [14, 15].

1.4   SOLAR PHOTOVOLTAIC POWER GENERATION

Solar cells, also known as PVs, are electronic devices that transform sunlight directly into electricity. The sun's energy can be considered the primary source of all forms of energy. It can be used in several ways, including using all of the sun's energy to produce electricity directly or using the sun's heat as a thermal energy source. In the solar energy area, PV cells are widely used. Experts forecast that, by the year 2040, PV systems will have the ability to produce over 72 TWh of electricity annually. Due to recent advances in the manufacturing process and lower production costs, new solar PV systems have the highest total generating capacity. There are three types of PV devices: crystalline, thin-film, and concentrated PV (CPV). They are all different.

1.4.1   SOLAR PV TECHNOLOGY TYPES

PVs typically require multifaceted materials that directly transform sunlight into DC electrical power. A PV effect is a critical mechanism for this energy conversion. Solar cells are usually configured as P-type and N-type semiconductors. Electrons are abundant on the N-type semiconductors' material on the one side, and the P-type semiconductor on the other side shows excess holes, each of which means an electron is absent.

The creation of electrical current in a solar cell, often called "light-generated current," is a two-step process. The first stage includes the absorption of the incoming photon, which results in the generation of an electro-hole pair. The second phase includes the acquisition of these carriers by the P-N junction, which prevents their recombination by physically isolating the electrons from the holes. An external circuit is used to enable the passage of charges, i.e., electrical current, through the system.

Alternatively, the PV cell generates a voltage through a development known as the "photovoltaic effect." A collection of light-generated carriers increases the number of electrons and holes on the N- and P-type sides, respectively, of the P-N junction. As a result, the junction's electric field is lowered, resulting in an increase in diffusion current. An equilibrium state is reached where voltage is developed across the P-N junction.

Solar PV modules are built by joining PV panels in series and parallel. A module is typically composed of 36 or 72 series-connected cells. PV modules are then connected in series to create strings, which are subsequently connected in parallel to create PV arrays, as depicted in Figure 1.10. This configuration of solar cells and modules enables the PV array's terminals to generate high current and voltage.

The solar cell's effectiveness has been found to range between 15% and 22%, with continuous improvements in the material of the semiconductor cells being pursued. Several decades of work, growth, testing, and trial and error have contributed to the large variety of solar panels in today's market. Several solar cells are available, and the development continues in this area to improve energy output efficiency.

PV devices are generally divided into the following three different types based on the type of the used material:

  i. **Crystalline silicon:** first-generation solar technology
 ii. **Thin-film:** second-generation technology
iii. **CPV:** third-generation solar technology

Although there are numerous other types of solar cells under each of the previous main classifications, the predominantly existing cells currently in the market are

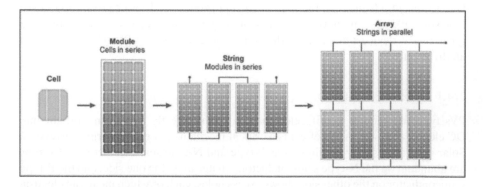

FIGURE 1.10   Configuration of solar cell, module, string, and array [16].

monocrystalline (Mono-Si), polycrystalline (p-Si), and amorphous silicon (A-Si) panels.

  i. **Monocrystalline**: A monocrystalline panel is a crystal panel composed of only one polycrystalline silicon-silicon ingot; hence, it produces only one monocrystalline crystal when manufactured (Mono-Si). This kind of silicon PV panel (monocrystalline) is the highest quality. It is easy to pick this panel out by its shape and rounded edges. Of the various silicon types, Mono-Si is the most used cell in commercial applications with efficiency reaching 20% with a high market cost. The panels that use monocrystalline need less volume and outlast monobloc ones. Also, as a result, they are often the costliest. They are less sensitive to changes in temperature than polystyrene panels. Monocrystalline panel types are manufactured by melting silicon that is less expensive and is time-consuming than refined. However, the distinctions between polycrystalline and monocrystalline forms are not enormous. Mono-Si provides marginally better room for quality at a higher price, with almost the same power output.
  ii. **Thin-film**: The second-generation type in PV technology, thin-film PV arrays, of various forms, are typically integrated into buildings or used in smaller installations. Thin-film solar cell technology is employed by covering a substrate with one or more layers of light-transmitting material (silicon, cadmium, or copper) with one or more films of light-sensitive material (such as a PV coating). These solar panels are much easier to make and more cost-effective to manufacture because they use less material and have lower costs.
  iii. **CPV**: It has a very high performance and efficiency rate compared with the others, reaching up to 41%; on the other hand, it needs a solar tracker and cooling arrangement (to achieve a high level of efficiency). Even while P-Si and A-Si cells are less expensive than other types, they have a shorter lifetime because they are more susceptible to high temperatures.

Because of the sharp decline in cost of production and technological advancements, experts forecast that, by the year 2050, PV systems would have the ability to produce over 9,667 TWh of electricity annually if stated policies are followed [17].

1.4.2 Solar PV in Oman

In 2017, to encourage the use of small-scale, grid-connected solar PV systems by consumers in Oman, the Authority for Public Services Regulation (APSR) updated the current regulatory regime. The PV guidelines detail the following:

  i. Establishing the minimum technical standards required for small-size grid-connected PV systems
  ii. Detailing the connection process including the installation, metering, and operation/maintenance

iii. Allowing distribution companies to behave as representatives on behalf of OPWP (the existing single buyer of electricity) to purchase PV-generated electricity from customers.

With the introduction of the Sahim-I and Sahim-II (approximately 3–5 kWp for residential rooftops) initiatives by the APSR in 2017 and 2019, respectively, there has been an increase in the uptake of small- to medium-grid interconnected PV systems. In accordance with Sahim-II, APSR anticipates that roughly 10% of residential premises will have rooftop PV systems, with private developers permitted to develop, manage, and operate PV systems at residential premises authorized by APSR [18].

As a result, distribution utilities in Oman anticipate problems and possibilities in their planning and operation of distribution systems. Although PV may have the effect of lowering transmission network congestion, it is important to note that if PV adoption becomes widespread, issues such as power quality, reverse energy flow, and running the network within acceptable voltage limits will need to be addressed by distribution utilities in Oman.

1.4.3 PHOTOVOLTAIC CONTROL REQUIREMENTS

Solar PV control systems have lately drawn a great deal of interest. Several control devices, controllers, and techniques have been widely published in the literature.

Inverters in the solar PV systems are the fundamental components because they perform the required control functions. Inverters convert DC voltage from PV output into AC waveforms for the electrical power grid, in grid-connected systems. At the PCC, the AC waveform must comply with the grid's amplitude and frequency standards. To preserve the relation, phase-locked loop (PLL) techniques or synchronization algorithms must be used to coordinate the inverter outputs with the grid voltages. Controlling the harmonic content injected at the PCC is also critical. Power management, monitoring and controlling, communication, heat monitoring, and protection are some of the frequent functions of the inverter in solar power systems. Independent and grid-connected networks need different levels of control. In Figure 1.11, power can be transferred to the load and supplied from PV panels under two system configurations:

i. An independent/islanded system
ii. Interconnected to the power grid

Maximum power and power efficiency are critical PV system control aims. An MPPT algorithm and a controller are needed in both situations. The MPPT method is used to maximize the amount of power that can be extracted from a solar panel. In independent power systems, extra controllers may be added to handle BESS and distribution functions for improving the delivered power efficiency.

The control circuit must contain anti-islanding safety, voltage/frequency ride-through, and control while simultaneously handling anti-island faults. If the power system complies with grid the code requirements of voltage, frequency,

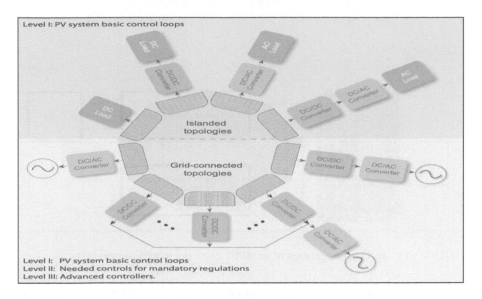

FIGURE 1.11   Array topologies and their control architectures [8].

reactive power, and power factor specifications must be part of the design parameters.

Most control mechanisms incorporate both linear and nonlinear techniques and classical and artificial approaches. Classical control techniques are combined with PV systems and electrical power converters.

1.4.4 MAXIMUM POWER POINT TRACKING METHODS

The I–V (current–voltage) characteristics of a PV array are nonlinear, its output power is strongly reliant on solar insolation and ambient temperature, and it fluctuates with these parameters as illustrated in Figure 1.12. Power–voltage (P–V) curves

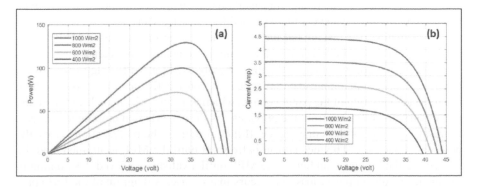

FIGURE 1.12   (a) P–V curves. (b) I–V curves, at 25°C, for varied irradiance at different temperatures [7].

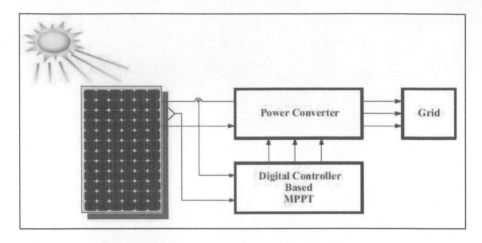

FIGURE 1.13   Block representation of an MPPT controller in a PV system [19].

have only one point, known as the maximum power point (MPP), where power is maximized. The location of this point changes with changes in ambient conditions. Also, the energy conversion efficiency of a solar module is quite low, and mismatches between the features of the source and the characteristics of the load result in significant power loss. As a result, it is vital to maximize power production while also enhancing efficiency [19].

The MPPT provides the maximum power for the PV system. To achieve MPPT, a power converter, typically a DC-DC converter in conjunction with an inverter, is employed between the PV module and the local load, as shown in Figure 1.13 [19].

If the PV voltage is increased or decreased using a boost converter, the MPPT controller changes the duty cycle to match the system output. The two values are then obtained and sampled and compared with the resultant current and voltage.

Literature review shows that there are numerous methods for achieving the MPPT and these methods are also classified in different ways as in "conventional methods" and more advanced "soft computing methods" as well as "online" and "off-line" methods, based on parameters that are used for tracking MPP.

In the context of conventional methodologies, the most often used MPPT techniques are perturb and observation (P&O) and incremental conductance [20].

The P&O approach, which is the most widely used for MPPT algorithms, is an iterative algorithm that depends a great deal on constructive feedback power. There is a repeated variation of the panel voltage, which is a comparison made of that to the power variation resulting from the previous. As the perturbation amplitude changes, it must be retained at the same degree. If the position of the perturbation varies, the power will decrease. The mechanism is frequent until the MPP is attained, and the device oscillates above and below the MPP until a value is achieved. Oscillation in the device results in power loss that occurs in any case. A diagram of the P&O process is shown in Figure 1.14.

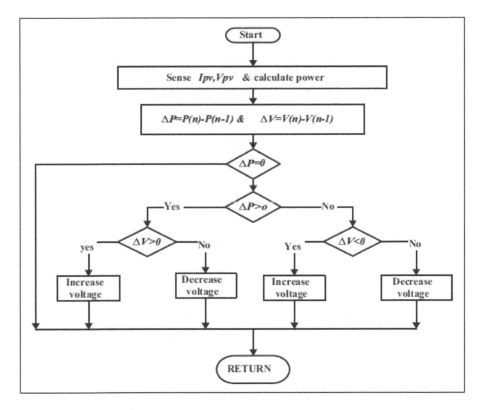

FIGURE 1.14   Perturb and observe MPPT algorithm [19].

1.4.5 Structure of Controller for Three-Phase Grid-Connected PV Inverter

The two main classes of inverters are named line-switched and self-switched. Additionally, inverters are also organized into a range of subcategories based on the types of source, modulation technique, output signal characteristics, network connection, and load. PV power injection (or feeding) must be constantly tracked and kept under control and analyzed. The grid-connected PV system architecture employs two control loops:

    i. Voltage control
    ii. Current control

The voltage control loop (or outer loop) regulates the contribution of real and reactive power from PV modules and balances the flow of current through the system, and current control regulates the flow of grid voltage to maintain the power factor.

    Today, many control systems have been suggested that control the grid injection. These controllers, such as the hysteresis controller, are predictive and linear

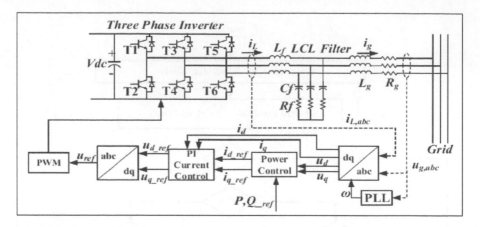

FIGURE 1.15   Schematic of a three-phase VSI [21].

proportional, PI, and exponential (EXP). A PI controller is the most used modern error-compensation management algorithm. A PI controller computes an error value as the variance between observed and expected output current into the grid and minimizes the difference between them. Typically, most controllers work in the rotating d-q axes reference frame.

A typical schematic of a three-phase on-grid voltage source inverter (VSI) is depicted in Figure 1.15. In VSI, the three-phase voltage and current values output by the VSI are translated to the revolving d-q axis frame. The reference values P_{ref} and Q_{ref} are used in the outer loop control to obtain the set point of current values $id_{\_ref}$ and $iq_{\_ref}$. These reference current values are then supplied into the inner control loop's PI controller and compared with the feedback values i_d and i_q to determine the requisite reference voltages $ud_{\_ref}$ and $uq_{\_ref}$. The reference regulating voltage is then converted back to the A, B, C format and delivered to the pulse-width modulation (PWM) controller in the final stage.

Because cross-coupling exists between the d and q mechanisms of the electric current in PV VSI-based control, it is necessary to reduce this cross-coupling to allow for individual control of both real and reactive powers.

1.5   BATTERY ENERGY STORAGE SYSTEM

The requirement for electricity storage devices and their application in power systems has been debated strongly for years. Despite the enormous number of studies that have been conducted on the application of various storage technologies in power systems, only a small number of them have found their place in the practical world.

As power systems experience a paradigm shift with PV and wind power generations expanding globally along with continual development and increasing deployment of MGs, the BESS provides promising options for increasing the stability and dependability of the electric power system. The ability of BESS to absorb and deliver both active and reactive power with subsecond response times makes them technically viable in offsetting the integration issues of the solar PV and wind generation such as ramp rate, frequency, and voltage. Furthermore, when compared with other energy storage technologies such as flywheels, pump-hydro, capacitors, supercapacitors, and compressed air energy storage, batteries have a higher capacity and greater siting flexibility than these technologies.

Researchers have studied in detail the services that BESS can offer including support for transient frequency stability [22], leveling of output power [23], transmission and distribution congestion management [24], control of ramp rate [25], peak load shaving [26], and network reliability enhancement [27]. A BESS is also sought to provide black start capability. The active power regulation of BESS supports the power system by improving the frequency regulation, output power variations, and dispatchability. Similarly, by regulating the reactive power of the BESS, it helps in reactive power reinforcement and voltage regulation [28]. Furthermore, the integration of BESS in RES-based generation sites is an enabler for solar generation to be more economical in energy markets.

Oman is set to witness the debut of BESS projects in the sultanate. The RAECO (Tanweer) and Petroleum Development Oman (PDO), as well as its parent company Energy Development Oman, have announced plans to establish a series of small-scale solar PV–diesel hybrid projects around the Sultanate of Oman.

Tanweer has invited proposals for 11 solar–PV and DG hybrid generation projects due to be developed at remote places across the utility's wide license. An aggregate of 28 MW of BESS is planned to be deployed across these projects to reduce the demand on the costly diesel-based generation while ensuring fast response in the occasion of supply shortfall or outage, which is a key requirement for the isolated grid operated by Tanweer [29].

On the other hand, PDO aims to establish a renewable energy independent power producer (IPP) project in their Block 6 concession in northern Oman. A large-scale battery storage component is envisaged for the 100-MW solar IPP project to ensure sustainable and reliable power supplies. The battery storage part is expected to be approximately 30 MW in capacity [30].

1.5.1 BESS-System Architecture

The corresponding circuit of a battery unit is shown in Figure 1.16. To keep the BESS running continually, the battery state of charge (SOC) must be kept within a specified range, which can consequently prevent the forced BESS shutdown due to battery overcharge or discharge. The SOC of a battery is the capability of the battery to engage in charging/discharging cycles with respect to the power and energy application. The adaptive coordination of the smoothing level and power distribution among energy storage units is considered in relation to the SOC and the highest allowable charge or discharge power capabilities of BESS, respectively [31].

Generally, the voltage at the battery terminals is expressed as shown in Equation 1.3. The battery's SOC is calculated using Equations 1.7 to 1.9. The "−" sign is used to denote the BESS in charging power condition while "+" indicates the discharging power of BESS [31]

$$V_{bat} = V_{ocv} - R_{intbat}I_{bat} \qquad (1.3)$$

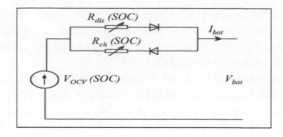

FIGURE 1.16   Equivalent circuit of the battery unit [31].

where:

$$V_{oc} = f_1(SOC) \tag{1.4}$$

$$R_{intbat} = R_{ch} = f_2(SOC) \text{ charging} \tag{1.5}$$

$$R_{intbat} = R_{dis} = f_3(SOC) \text{ discharging} \tag{1.6}$$

$$SOC = SOC_{ini} - \int \frac{\eta I_{bat}}{Q_{bat}} dt \tag{1.7}$$

$$\eta_{ch} = \frac{V_{ocv}}{V_{ocv} - I_{bat}R_{Rch}} \text{Charging} \tag{1.8}$$

$$\eta_{dis} = \frac{V_{ocv} - I_{bat}R_{Rch}}{V_{ocv}} \text{Discharging} \tag{1.9}$$

A grid-scale BESS is made up of several components, including batteries, a control and power conversion and conditioning system that is made up of power electronics for AC-DC power conversion, associated protection systems, and a transformer to convert the output to distribution or transmission voltages. A bidirectional inverter is the key element that performs power conversion between the AC line voltage and the DC battery terminals and permits the flow of power in both charging and discharging modes of the battery.

A simple wind–solar PV–BESS hybrid scheme is shown in Figure 1.17.

1.5.2   BATTERY TYPES

According to [28], various different types of battery technologies employed in an RES-based MG are included in the following list:

  i. Lead acid
 ii. Lithium-ion (Li-ion)
iii. Sodium sulfur
 iv. Nickel cadmium
  v. Zinc hybrid cathode

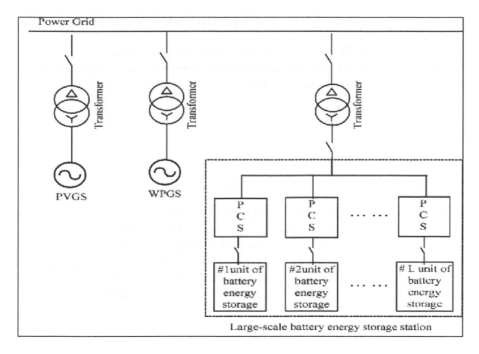

FIGURE 1.17 Wind-PV-BESS generation scheme [31]. PCS, power conversion system.

   vi. Vanadium redox battery
  vii. Polysulfide bromide
 viii. Zinc bromine

These battery types differ from each other in various important aspects like efficiency, energy and power density, cost, response time, and life cycle span, which define their merit to be suited for power system application. Some are better in terms of efficiency and power density, whereas others offer cost economy. For example, the Li-ion battery has an efficiency of nearly 100% compared with lead acid (70%–90%) or sodium sulfur (70%–90%) types, but it comes at a higher capital cost in terms of $/kWh. Zinc hybrid cathode, on the other hand, is nearly 50% cheaper than Li-ion at a price of about $160/kWh.

1.5.3 Power Converter System

The power conversion system (PCS) is a power electronics-based storage system interface. Whereas a BESS employs a DC interface, the battery and solar PV can share the same bus. Wind power terminals are not equipped to handle this. PCS is therefore required to connect a BESS to the AC grid.

    For the current MG, the PCS offers immediate active and reactive power control. The PCS has two levels of control: main and secondary. Depending on the system's status and reference charging mode, the primary controller creates gate drive signals. In response to SOC, electricity pricing, etc., secondary control selects appropriate operation mode.

Primary Control: The simplest control strategy is the PI control.

Secondary Control: This establishes the power converters' operating mode. Three frequently used techniques are as follows:

i. Charge mode
ii. Mode of discharge
iii. nactive mode

1.5.4   FREQUENCY REGULATION APPLICATION OF BESS

In this application, to manage the active power shortfall, PV and BESS, for example, can work in combination for frequency regulation. The features of the power frequency droop (P-f) as defined in Equation 1.10 can be fixed [32] or adaptable; however, an adaptive P-f exhibits smoother transitions in various control schemes [33]. As shown in Equation 1.10, according to fluctuations in frequency deviation from the nominal set point, the amount of active power regulation provided by BESS is determined by droop value RP-f.

$$P = \frac{df}{Rp - f} \tag{1.10}$$

When using a BESS, the frequency drop and the frequency oscillation are reduced.

1.5.5   KEY CHALLENGES IN BESS APPLICATION AND FUTURE DIRECTIVES

The review of the existing literature signifies the technical benefits of the BESS application in MGs in standings of providing reactive and active power support along with frequency regulation and other potential benefits as discussed previously. Discounting the fact for the various developments and application areas being pursued currently for the BESS, a wide number of opportunities are present for further research in this field. A few of the present-day key challenges as summarized in [28] include the following:

i. An important consideration in BESS sizing is the battery's energy efficiency and longevity. Thus, it is essential to enhance battery technology to lower battery costs.
ii. The costs of BESS installation have so far been prohibitive to the adoption of BESS by RES farm owners. This means that BESS installation must be facilitated by a cost-effective and efficient BESS technology.
iii. The impact assessment of battery storage system failures and disturbances on the associated network is open for study
iv. Often the dynamic influence of battery efficiency on grid support is overlooked. Conversion efficiency, including variable charging/discharging resistance, must be considered in real time in relation to battery SOC.

v. Accurate forecasting enables better unit commitment planning by efficient and economical management of the battery charging/discharging. However, reliable prediction is frequently challenging; hence, progress in estimating irradiance and wind speed is critical.

1.6 TYPES OF CONTROL FOR MICROGRID

The control systems of MGs must regulate voltage and frequency precisely in grid-connected and island mode while staying within defined limitations for power quality and reliability [14]. Detailed literature review in this area informs that MG control has come out to be the most challenging area of research in the field of MGs. A great deal of research has been done and is continually progressing for the development of precise control systems and techniques to ensure perfect operations of AC MGs in challenging network conditions.

1.6.1 Modes of Operation of Microgrid

One of the two possible operation modes for an MG is as follows:

i. **Grid connected mode:** In this mode of operation, the MG is connected to the power grid via a static transfer switch. Also known as "point of common coupling," this location serves as a point of connection. Because of the inverter's constant monitoring of power generation and consumption inside the MG, surplus or deficit electricity can be exported or imported based on load and source conditions, respectively. Once it is connected to the grid, an MG will no longer be able to regulate the system's frequency or voltage, and it will, instead, move to P-Q control for real and reactive power regulation [34].

ii. **Islanded:** In islanded operation mode, there is no grid support and MG control becomes more complex. The MG's low inertia makes it extremely sensitive to changes in generation and demand. An island MG relies on a steady supply of electricity. Devices for storing electrical energy are commonly used [35]. Controlling storage unit devices like batteries and super capacitors efficiently keeps the voltage and frequency constant [36], and DG, by running in isochronous mode, helps in maintaining the inertia and hence the frequency [14, 15].

Master-slave control, communication via control area networks, and voltage and frequency droop schemes based on local measurements have all been investigated in islanded MGs [37]. It is possible to use centralized or decentralized control in an autonomous mode to set the voltage and frequency of the system.

1.6.2 Control Techniques in Microgrid

Methods of control vary in difficulty and level of performance, speed, and accuracy. A variety of control methods are employed in AC MG control:

i. **P-Q control:** This method is used in grid-connected operation mode; hence, the MG controller is not required to follow the V-f control method. To achieve economic operation, the inverters now inject maximum power into the grid.

ii. **V-f control:** This method is used in islanded mode; hence, the MG controller must meet the requirements of the connected load while maintaining the frequency. Because a pure V-f mode cannot effectively respond to load fluctuations, master-slave control configurations are proposed [38].

iii. **Droop control:** It is necessary to use conventional voltage and frequency drooping characteristics for the inverters of an MG, which means that the active output power and real and reactive absorbed power are inversely related to the frequency and terminal voltage, respectively. The electricity demand is shared among the power-producing units according to the droop characteristic functions of each unit in this technique.

1.6.3   CONTROL ASPECTS OF THE MICROGRID

Before describing the types of AC MG control, it is important to be briefed on the control strategies based on the hierarchical architecture Reference [39] clearly articulates the various MG control aspects of MGs as summarized in the following:

i. **Inner and outer loop control:** The internal loop control is responsible for controlling the source operating point, i.e., for controlling the output network current and managing the output powers of RES. Power electronics are used for this functionality, hence, this is also called the converter output control.

ii. **Primary control level (or power sharing control layer):** At this level, the voltage and frequency of the inner loop are regulated, and the same reference is provided to both loops. Droop control is most commonly employed at this level.

iii. **Supervisory control layer:** MG output powers are controlled by this mechanism, which restores voltage and frequency to their pre-interference state. Depending on the method of operation, the supervisory control level can be further subdivided into two levels:

     a. **Secondary control level (or MG supervisory control):** Secondary control, also defined as load frequency control (LFC), regulates an MG's frequency and voltage even after a fault has occurred. LFC is also responsible for power exchanges between several control zones. MGs with low-bandwidth communications and slower control loops have secondary control at the top of the hierarchical structure in island mode.

     b. In addition, secondary control is classified into "centralized control" and "decentralized control" depending on the position of the MG central controller in the system.

     c. **Tertiary control level (or grid supervisory control):** In grid-connected mode, the MG's internal frequency and voltage are controlled

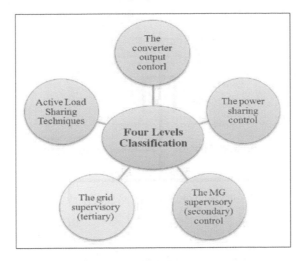

FIGURE 1.18   Four levels of AC MG control strategies [39].

by tertiary control to regulate power flow between the MG and the grid at the PCC. For this reason, when using islanded mode, the references for the tertiary control must be disconnected [37, 39].

Figure 1.18 captures the various MG control strategies.

1.6.4 CENTRALIZED CONTROL

When there is a fixed infrastructure and separated MGs with demand-supply balances, central controllers are more suitable [39]. In a centralized control system, load measurement, such as current, voltage, or power, for each unit, measurement is sent to a single point where control signals are generated for the whole system. This kind of control necessitates a network connection between the power plant and the source unit to collect input and give a signal to all the new conditions. However, having a communication link means an extra expense. Also, the loss of coordination will cause the process to fail.

In recent years, the principle of centralized control has been implemented to MGs to recover the nominal values of frequency and voltage after years of use in large utility power grids, as proven in a number of studies [40, 41]. The block representation of a centralized secondary control level is shown in Figure 1.19.

1.6.5 DECENTRALIZED CONTROL

A decentralized controller should be employed to maximize the independence of multiple generation sources like wind, solar PV, BESS, and diesel engine generators to address the MG's energy management challenge. According to the concept of decentralized control, a complicated wide-area power system must be separated into subsystems, with each subsystem controlled by a different controller [42]. Multiple

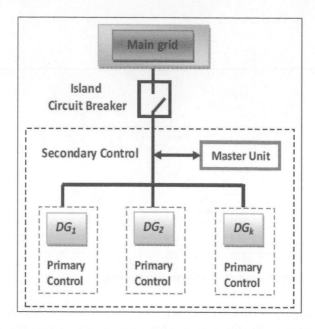

FIGURE 1.19   Block diagram of centralized type of secondary control [39].

controllers can therefore be used for droop performance. For frequency and voltage restoration, distributed control techniques such as multiagent systems (MASs) can be used, even if the secondary layer of MG is designated the primary controller [39].

Due to the distributed nature of the system as the generation sources are dispersed within the MG, the interfaces must be regulated solely by local measurements, i.e., data transfer is not desirable. The following fundamental difficulties as mentioned in [43] should be addressed by the decentralized control of the specific interfaces.

i. The interfaces should be configured to distribute the overall load in the desired manner.
ii. Ensure global stability based on local measurements.
iii. The inverter controller must avoid any DC voltage offsets on the MG system.
iv. Active damping of oscillations between the output filters should be accomplished via the inverter control.

From the standpoint of decentralized control, the MG generation architecture can be classified into the following:

Highly dispersed network: This is where the impedances are mainly inductive and the voltage amplitude and phase angle at various source interconnects can be substantially different.

Smaller network area: While still being inductive, the impedances include a large resistive component as well. However, phase angles might differ from one device to another.

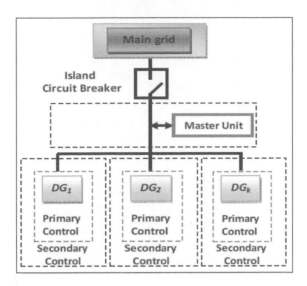

FIGURE 1.20 Decentralized type of secondary control [39].

Very small network area: The impedance of small network is low and mostly resistive. At no time are magnitude or phase angle discrepancies substantial. The steady-state frequency, which must be the same for all sources, is the most important constant under all conditions.

The block representation of a decentralized secondary control level is shown in Figure 1.20.

It is confirmed that decentralized methods are superior to a centralized controller; however, notwithstanding their advantages, when applied to LFC, decentralized methods have the following shortcomings as pointed out in [42] and mentioned in Section 1.7.

Combining items from the previous discussion on the various control techniques and control aspects of the MG, the overall hierarchical control of an MG is presented in Figure 1.21.

1.7 NEED FOR FURTHER WORKS IN FREQUENCY CONTROL

From literature review, it becomes clear that although so much research is done and underway about MGs and the frequency control, there is still a need for further deeper investigations around voltage and frequency control methods, and making them fully developed, field demonstrated, and thoroughly experimented for islanded and grid connected operation modes cannot be discounted. Summarizing the findings of [42], which has comprehensively reviewed the subject of LFC covering over 54 years of published research, it is evident that future power systems will experience huge challenges around frequency control. Modern power systems will be deficient in system inertia and suitable damping. This necessitates the need for developing highly precise and suitable control mechanisms and secondary operations like

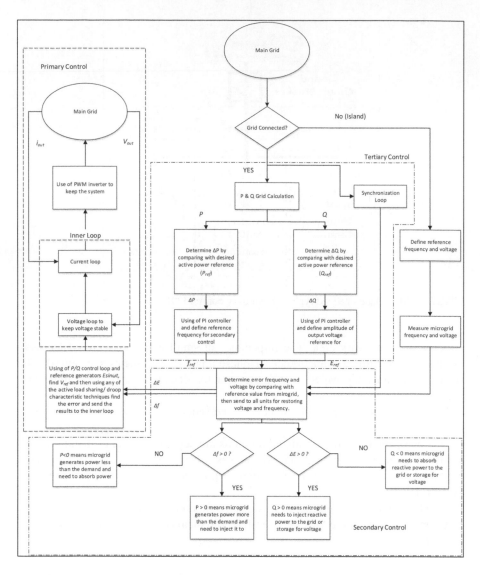

FIGURE 1.21 Hierarchical control of a an MG [37].

main and backup reserves for constant monitoring and rapid control of frequency. The literature survey highlights research limitations in the topic of LFC [42], which will potentially require further work:

i. Enhancing the robustness of LFC control mechanisms.
ii. Proposing new techniques of control by making use of a wide-area monitoring system.
iii. Detailed investigation of the reliability aspects of LFC loops.
iv. Creating power system control techniques that don't rely on assumptions.

v. Ensuring that the LFC system can handle cyberattacks more effectively.
vi. Propose appropriate control algorithms capable of detecting and isolating sensor failures in LFC loops.
vii. The interface between the LFC as well as other control loops, including the LFC and automatic voltage regulator (AVR) control loops, should be studied.
viii. New methods of flaw detection that are appropriate for LFC are required.

Adequate development and field testing of ancillary services for virtual inertia, using improved data from PV and wind forecast in inertia estimation, is required to improve frequency stability.

ix. A discussion on system-level integration of virtual inertia systems for frequency stability in future power systems in [44] points out several areas potentially requiring research to answer critical questions.
x. Detailed impact assessment and real-time testing of the interaction between the virtual inertia systems and existing synchronous machines is a topic of critical importance for future research.
xi. As the number of virtual inertia systems grows in the future, the coordinated and aggregated operation, as well as the optimal placement of these systems, will potentially become critical questions requiring research attention.
xii. Because the current literature is deficient in the availability of accurate mathematical models to represent the dynamics of the virtual inertia systems, there is a need for the development of the same as these models will be highly essential for parameter optimization and understanding the working behaviors as these systems are interconnected to the main power system.

1.8 HARDWARE IN THE LOOP

There has been a high interest in real-time modeling and simulation developed in recent years to study the complex nature of operation and control of power systems, for example, MGs and their control systems. Real-time simulation allows for design changes to be implemented earlier in the process, hence, lowering costs and shortening the design cycle. Due to these factors researchers in their studies for regulation, management, and security assessment of MGs have typically used real-time and offline simulations in the early stages of their study.

1.8.1 DIGITAL REAL-TIME SIMULATIONS

Digital real-time simulation (DRTS) of the electric power system is the process of reproducing output (voltage/current) waveforms with the required accuracy that is reflective of the real power system being modeled.

DRTS is a method for transient simulation of the power systems that makes use of a time-domain solution on a digital computer (e.g., employing a transient electromagnetic technique). Systems are defined using the element models from the

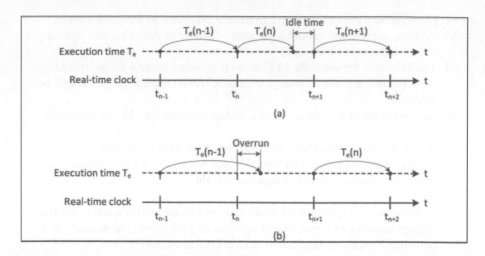

FIGURE 1.22 A visual representation of both real-time and off-line simulation. (a) Simulating in real time. (b) Simulating in non-real time [45].

software tool's library and are then simulated utilizing parallel processing on a hardware platform [45].

Depending on the simulation platform's time required to compute state outputs for each time step, two scenarios can occur. The simulation is regarded as real time if the implementation time, T_e, is less than or equal to the set time step; otherwise, non–real-time or off-line simulations are used when time constraints are exceeded as shown in Figure 1.22, for which case a higher time step or a simpler scheme model can be used to run in real time as a solution [45].

The DRTS can be broadly classified into two categories for power systems simulations.

  i. **Purely software or fully DRTS:** This requires modeling of the full system in software and doesn't include external I/Os for interfacing.
 ii. **Co-simulation (software and hardware) or hardware-in the-loop (HIL) real-time simulation:** This denotes a condition in which actual physical components replace some parts of the DRTS completely.

Figure 1.23 shows the further classification of DRTS.

1.8.2 REAL-TIME SIMULATORS

Numerous digital real-time simulators (RTs) were proposed and tested employing parallel processor-based computer software and improved numerical analysis approaches that require less computing capabilities [46–48].

There are several RT platforms available in the market today. Utility companies have conducted closed-loop testing using the real-time digital simulator (RTDS),

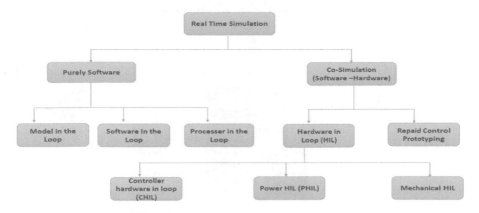

FIGURE 1.23 Classification of real-time simulations.

large-scale tests, and controller testing of protective relays for many years. As articulated in the following list, the use of RTs in some of the recent studies proves that it is the most convenient approach for studying MGs.

i. **RTDS** was utilized to develop an RT hardware test platform for analyzing the system response of a simulated power model using phasor measurement unit (PMU). Also, representing an MG modeling platform based on IEC61850 and NIcRIO is used to validate a potential power control method in multi-RESs. Moreover, the system is utilized to validate the control method for a hybrid energy system in an MG. Additionally, MG control algorithms and management strategies will be evaluated [49–53].

ii. **The laboratory prototype** is described as a grid test [9] that emulates a DFIG and permanent magnet synchronous generation and voltage and frequency testing of the proposed RT strategy [54, 55].

iii. **OPAL RT** is used to operate the DC MG's controller, testing, and operation. Evaluating the performance of a control mechanism in a hybrid power system with numerous subgrids [56–58].

Figure 1.24 shows an overview of MG simulation concept in RTDS.

1.8.3 Hardware-in-the-Loop Simulation

For the development and testing of complicated, real-time embedded systems, HIL simulation (HILS) is a technique that uses electrical equivalent representations of sensors and actuators.

These electrical emulations allow the plant model and the embedded systems under evaluation to interact with one other. Control of sensor values is performed by the plant model, which is subsequently read by the embedded system under test. Control signals are generated for actuators as part of the embedded system's testing of its control algorithms. Control signals affect the values of variables in the plant

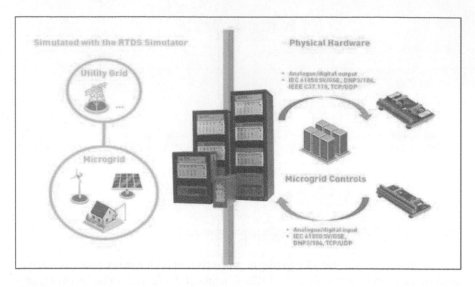

FIGURE 1.24 MG simulation concept in RTDS.

simulation [49]. Figure 1.25 shows the generic conceptual structure of the HILS system.

However, using existing HIL setups for academic research is still costly. This research has therefore undertaken a rather low-cost HIL setup using commercial off-the-shelf (COTS) devices commonly found in university laboratories such as MATLAB® to simulate and test the proposed MG and its hierarchical control in real time.

A HIL framework is therefore designed as seen in Figure 1.26, which models the MG in RTDS and the central controller in Simulink Desktop Real-Time (SLDRT). The methodology section explains in detail the proposed arrangement.

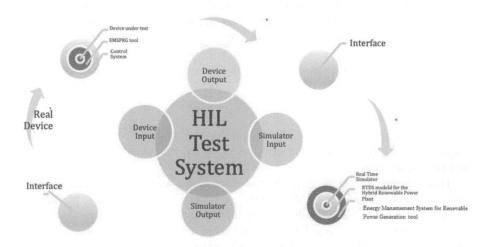

FIGURE 1.25 The conceptual structure of the HILS system.

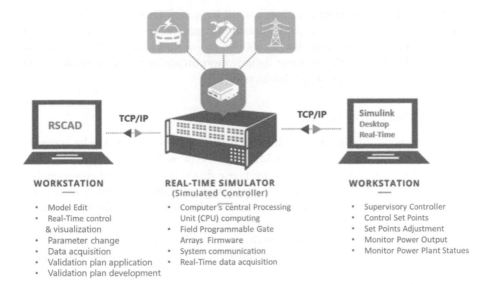

WORKSTATION

- Model Edit
- Real-Time control & visualization
- Parameter change
- Data acquisition
- Validation plan application
- Validation plan development

REAL-TIME SIMULATOR
(Simulated Controller)

- Computer's central Processing Unit (CPU) computing
- Field Programmable Gate Arrays Firmware
- System communication
- Real-Time data acquisition

WORKSTATION

- Supervisory Controller
- Control Set Points
- Set Points Adjustment
- Monitor Power Output
- Monitor Power Plant Statues

FIGURE 1.26 The HIL framework for this research.

1.9 OPTIMIZATION

The electric power business is experiencing extraordinary restructuring presently. The deregulation of the power supply business has created new opportunities for the race to reduce costs and lower prices. Maintaining an economical and dependable supply of power in such a landscape is a huge concern for utilities. Facing the challenge of an increasingly difficult existence, the utilities require effective tools and assistance to ensure that the desired quality of electrical energy is given at the lowest possible cost. For both short-term and long-term operations, the primary objective is to achieve the best possible balance between security and economic needs. To put it another way, it is fundamental to have effective tools for solving complex optimization problems [59].

The usage of distributed computing has been one of the significant modem optimization technologies that have been used to power systems in recent years. There has been a vast range of research and reports. To assist the difficult task of supplying electricity to the grid in an efficient and dependable way, the use of optimization technologies in power system regions is becoming increasingly important. The use of energy management, power system operation, analysis, and schedule optimization approaches is necessary in a variety of power system contexts. Different approaches to solving the challenges in these domains are required, depending on the objective function and constraints being studied [60].

Nearly all fields of research, engineering, and business are affected by the global optimization problem in some way. The process of determining the optimal parameter configuration falls under the area of global optimization. Generally, the principal aim of optimization methods is to keep the system's development and operation costs to a minimum. As a result, the objective function is economic efficiency or profit, whereas the restrictions represent the other needs. Long-term, medium-term, short-term, and even online decisions can all be made. There may be a hierarchical framework that includes expansion planning, maintenance scheduling, and fuel resource scheduling to ensure that the optimal power flow may be achieved. A new dimension has been added to these optimization difficulties in light of the recent deregulation of the energy supply business [59].

1.9.1 OPTIMIZATION PROBLEM FOR THIS RESEARCH

Controlling voltage frequency is critical in power systems that need to be correctly regulated. It's challenging to operate MGs because of problems such as voltage collapses, voltage and frequency variations, phase difference faults, and power quality issues. MG stability is essential for the provision of transient stability employing intelligent optimization methodologies to tackle the challenges outlined previously that have a negative effect on power quality.

The ability of a power system to maintain a stable frequency following a major system upset resulting in a significant imbalance between generation and load is defined as "frequency stability."

When a disturbance occurs, load frequency management techniques should accomplish two primary goals:

  i. Restoring the frequency to its desired value
 ii. Maintaining the interchange power of tie-lines at the levels that have been established

As a result, LFC should be adjusted ideally to accomplish these goals.

1.9.2 OPTIMIZATION ALGORITHM FOR THIS PROBLEM

Although several optimization algorithms have been studied by researchers around the world in connection to the frequency stability in power systems, the imperialist competitive algorithm optimization is employed to optimize the gain controller to maximize the LFC in this research due to its proven advantages over others.

1.10 SUMMARY

This chapter presents the theoretical aspects of MGs and different components related to the MG's frequency control. Initially, the chapter began by presenting the most difficult aspects of frequency control in MG systems, such as the uncertainty of the RES in either grid-connected or islanded operation mode. Also, the chapter presents all the background and the concept of each unit source in (PV, wind, DG, and BESS). In addition, in each section of the unit source, a power generation control type and method are described. Moreover, to match the chapter's objective, the most well-known type of MG control has been discussed and reviewed. The review explained the differences between the types and the views of other projects for each type. As a summary of the other work, a list of future work in frequency control has been discussed. Finally, in line with the chapter objective, HIL has been explored with optimization technology.

REFERENCES

[1] Baggini, A. (2008). *Handbook of Power Quality*, John Wiley & Sons, 2008. The Atrium, Southern Gate, Chichester, West Sussex PO19 8SQ, England, ISBN-10:0470754230, Pages: 642.

[2] Madureira, A., Moreira, C. & Peças Lopes, J. (2005). Secondary load-frequency control for microgrids in islanded operation. *RE&PQJ, 1*(3), 482–486.

[3] Yasin, A., Napoli, G. Ferraro, M. & Antonucci, V. (2011). Modelling and control of a residential wind/PV/battery hybrid power system with performance analysis. *Journal of Applied Sciences, 11*, 3663–3676.

[4] Abdelrazek, S. A. (2016). Integrated PV capacity firming and energy time shift battery energy storage management using energy-oriented optimization. *IEEE Transactions on Industry Applications, 52*(3), 2607–2617.

[5] Abed Alaa, Kasim Naseer & Hussain Hazim (2020). Performance Improvement of CIGS PV Solar Grid Tied System Using Planer Concentrators, *Case Study: Baghdad*. doi:10.13140/RG.2.2.33172.73604

[6] Lasseter, R. et al. (April 2002). The CERTS Microgrid Concept. *White paper on Integration of Distributed Energy Resources*.

[7] Karaboğa, D. & Okdem, S. (2004). A simple and global optimization algorithm for engineering problems: Differential evolution algorithm. *Turkish Journal of Electrical Engineering and Computer Sciences, 12*, 53–60.

[8] Guerrero, C. A. V. et al. (2016). A new software-in-the-loop strategy for real-time testing of a coordinated Volt/Var Control. *IEEE PES PowerAfrica Conf. PowerAfrica*, 6–10. doi:10.1109/PowerAfrica.2016.7556559

[9] Omar Faruque, M. D. et al. (2015, June). Real-time simulation technologies for power systems design, testing, and analysis. *IEEE Power and Energy Technology Systems Journal, 2*(2), 63–73. doi:10.1109/JPETS.2015.2427370

[10] Alfergani, A., Alfaitori, K. A., Khalil, A. & Buaossa, N. (2018). Control strategies in AC microgrid: A brief review. *The 9th International Renewable Energy Congress (IREC 2018)* doi:10.1109/IREC.2018.8362575

[11] Atif, A. & Khalid, M. (2020). Saviztky–Golay filtering for solar power smoothing and ramp rate reduction based on controlled battery energy storage. *IEEE Access*, 33806–33817.

[12] (IRENA), International Renewable Energy Agency. (2019). *Renewable Energy Statistics 2019*. Abu Dhabi: www.irena.org/Publications.

[13] Bevrani, H. F. (2016). Robust frequency control in an Islanded microgrid: H∞ and μ-synthesis approaches. *IEEE Transactions on Smart Grid, 7*(2), 706–717. doi:10.1109/TSG.2015.2446984

[14] Bevrani, H. (n.d.). *Robust Power System Frequency Control*. University of Kurdistan, Sanandaj, Kurdistan, Iran: Springer Cham Heidelberg New York Dordrecht London.

[15] Bullich-Massagué, E., Aragüés-Peñalba, M., Sumper, A. & Boix-Aragones, O. (2017). Active power control in a hybrid PV-storage power plant for frequency support. *Solar Energy, 144*, 49–62.

[16] Busaidi, M. I. (2021). Design and Simulation of Dispatchable Hybrid Wind-PV Power Plant Control Scheme. *thesis*.

[17] Chien-Liang Chen, Yubin Wang, Jih-Sheng Lai, Yuang Shung Lee & Daniel Martin (2010). Design of parallel inverters for smooth mode transfer microgrid applications. *IEEE Trans Power Electron, 25*(1), 6–15.

[18] Moutis, P., Papathanassiou, S. A. & Hatziargyriou, N. D. (23 February 2012). Improved load-frequency control contribution of variable speed variable pitch wind generators. *School of Electrical and Computer Engineering, National Technical University of Athens, 9 Heroon Polytechniou St., Zografou, 15780 Athens, Greece*.

[19] Datta, U. K. (2017). Battery energy storage system for transient frequency stability enhancement of a large-scale power system. *Australasian Universities Power Engineering Conference (AUPEC)*, pp. 1–5.

[20] Atashpaz-Gargari, E. & Lucas, C. (2007). Imperialist competitive algorithm: an algorithm for optimization inspired by imperialistic competition. *Evolutionary Computation*, 4661–4667.

[21] Eduard, E. A. (July 2008). Wind power plant representation in large-scale power flow simulations in WECC. *Proc. 2008 IEEE PES General Meeting.*, pp. 1–6.

[22] Eltigani, D. & Masri, S. (2015). Challenges of integrating renewable energy sources to smart grids: A review. *Renewable and Sustainable Energy Review*, *52*, 770–780.

[23] Etemadi, A. H., Davison, E. J. & Iravani, R. (2012). A decentralized robust control strategy for multi-DER microgrids—Part II: performance evaluation. *IEEE Transactions on Power Delivery*, *27*(4), 1854–1862.

[24] Etxeberria-Otadui, V. M. (2002). Generalized average modelling of FACTS for real time simulation in ARENE. Proc. 28th Annu. Conf. IEEE Ind. Electron. Soc. (IECON), 2, 864–869.

[25] Chen, Z., Guerrero, J. M. & Blaabjerg, F. (2009, Aug). A review of the state of the art of power electronics for wind turbines. *IEEE Transactions on Power Electron*, *24*(8), 1859–1875.

[26] Mallesham, G., Mishra, S. & Jha, A. N. (2011). Ziegler-Nichols based controller parameters tuning for load frequency control in a microgrid. *International Conference on Energy, Automation and Signal*, 1–8. doi:10.1109/ICEAS.2011.6147128

[27] Guerrero, J. M., Chandorkar, M., Lee, T.-L. & Loh, P. C. (2013). Advanced control architectures for intelligent microgrids, Part I. *IEEE Transactions on Industrial Electronics*. doi:10.1109/TIE.2012.2194969

[28] Hassan Haes Alhelou, Mohamad-Esmail Hamedani-Golshan, Reza Zamani, Ehsan Heydarian-Forushani & Pierluigi Siano (2018). Challenges and opportunities of load frequency control in conventional, modern and future smart power systems: A comprehensive review. *Energies*, doi:10.3390/en11102497

[29] Software-in-the-loop | SIL testing | SIL software. (n.d.). https://www.opal-rt.com/software-in-the-loop/ (accessed Nov. 13, 2020).

[30] Tani, A., Camara, M. B. & Dakyo, B. (March/April 2015). Energy management in the decentralized generation systems based on renewable energy—Ultracapacitors and battery to compensate the wind/load power fluctuations. *IEEE Transactions on Industry Applications*, *51*(2), 1817–1827.

[31] Hongmei Tian, Fernando Mancilla-David, Kevin Ellis, Eduard Muljadi & Peter Jenkins (September 2012). A cell-to-module-to-array detailed model for photovoltaic panels. *Solar Energy*, *86*(9), 2695–2706.

[32] Hossam-Eldin, A., Refaey, M. & Farghly, A. (2015). A Review on Photovoltaic Solar Energy Technology and its Efficiency. *17th International Middle-East Power System Conference (MEPCON'15). December 2015*

[33] IEA (2021). *World Energy Outlook 2021, IEA, Paris.* Retrieved November 17, 2021, from https://www.iea.org/reports/world-energy-outlook

[34] IRENA (2020). *Renewable capacity statistics 2020 International Renewable Energy Agency.* www.irena.org/Publications.

[35] IRENA (2021, June). *IRENA_Power_Generation_Costs_2020.* (International Renewable Energy Agency) Retrieved November 16, 2021, from https://www.irena.org/costs/Power-Generation-Costs/Wind-Power

[36] Jeremy Lin, Fernando Magnago & Juan Manuel Alemany (2018). *Classical and Recent Aspects of Power System Optimization.* Academic Press. doi:https://doi.org/10.1016/B978-0-12-812441-3.00001-X

[37] Jin-Hong Jeon, Jong-Yul Kim, Hak-Man Kim, Seul-Ki Kim, Changhee Cho, Jang-Mok Kim, Jong-Bo Ahn & Kee-Young Nam (2010). Development of hardware in-the-loop simulation system for testing operation and control functions of microgrid. *IEEE Transactions on Power Electronics*, 25(12), 2919–2929.

[38] Jordehi, A. R. (2016). Optimal allocation of FACTS devices for static security enhancement in power systems via imperialistic competitive algorithm (ICA). *Applied Soft Computing*, 48, 317–328.

[39] Rajesh, K. S., Dash, S. S., Rajagopal, R. & Sridhar, R. (2017). A review on control of ac microgrid. *Renewable and Sustainable Energy Reviews*, 71, 814–819. doi:10.1016/j.rser.2016.12.106

[40] Keyhani, A. (2011). *Design of Smart Power Grid Renewable Energy Systems*. John Wiley & Sons, Inc., Publication.

[41] Le Luo, Lan Gao & Hehe Fu (2011). The control and modeling of diesel generator set in electric propulsion ship. *I.J. Information Technology and Computer Science*, 2, 31–37.

[42] Li, X. D. (2013). Battery energy storage station (BESS)-based smoothing control of photovoltaic (PV) and wind power generation fluctuations. *IEEE Transactions on Sustainable Energy*, 4, 464–473. doi:10.1109/TSTE.2013.2247428

[43] Liang, C. W. (2017). Battery energy storage selection based on a novel intermittent wind speed model for improving power system dynamic reliability. *IEEE Transactions on Smart Grid*, 9(6), 6084–6094.

[44] Bani Salim, M., Hayajneh, H. S., Mohammed, A. & Ozcelik, S. (2019). Robust direct adaptive controller design for photovoltaic maximum power point tracking application. *Energies*, 12(16). doi:10.3390/en12163182

[45] Moradi, M. H., Zeinalzadeh, A., Mohammadi, Y. & Abedini, M. (2014). An efficient hybrid method for solving the optimal sitting and sizing problem of DG and shunt capacitor banks simultaneously based on imperialist competitive algorithm and genetic algorithm. *International Journal of Electrical Power & Energy System*, 54, 101–111.

[46] Magro, M. C., Giannettoni, M. & Pinceti, P. (2018). Real time simulator for microgrids. *Electric Power Systems Research*, 160, 381–396.

[47] Mahmood, H. & Jiang, J. (2017). Decentralized power management of multiple PV, battery, and droop units in an islanded microgrid. *IEEE Transactions on Smart Grid*, 10(2), 1898–1906.

[48] Manbachi, M., Sadu, A., Farhangi, H. & Monti, A. (2016). Real-time co-simulation platform for smart grid volt-VAR optimization using IEC 61850. *IEEE Transactions on Industrial Informatics*, 12(4), 1392–1403.

[49] Merabet, A., Ahmed, K. T. & Ibrahim, H. (2017). Energy management and control system for laboratory scale microgrid based wind-PV-battery. *IEEE Transactions on Sustainable Energy*, 8(1), 145–154.

[50] Mohammad Reza Aghamohammadi & Hajar Abdolahinia (2016). A new approach for optimal sizing of battery energy storage system for primary frequency control of islanded microgrid. *International Journal of Electrical Power & Energy Systems*. www.elsevier.com/locate/.

[51] Muhammed Y. Worku, Mohamed A. Hassan & Mohamed A. Abido (2020). Real time-based under frequency control and energy management of microgrids. *Electronics*, 9(9). doi:10.3390/electronics9091487

[52] National Centre for Statistics and Information in Oman (n.d.). Retrieved from https://data.gov.om/bixytwb/weather

[53] Office, Oman 2040 Vision. (2019). *Oman 2040 Vision Document, 2019*. Retrieved from www.2040.om/wp-content/uploads/2020/12/Oman2040-En.pdf

[54] Lehn, P., Rittiger J. & Kulicke, B. (1995). Comparison of the ATP version of the EMTP and the NETOMAC program for simulation of HVDC systems. *IEEE Transactions on Power Delivery*, 10(4), 2048–2053.

[55] Bhatnagar, P. & Nema, R. K. (2013). Maximum power point tracking control techniques: State-of-the-art in photovoltaic applications. *Renewable and Sustainable Energy Reviews*, *23*, 224–241. doi:doi.org/10.1016/j.rser.2013.02.011

[56] Piagi, P. & Lasseter, R. H. (2006). Autonomous control of microgrids. *IEEE PES Meeting*. Montreal.

[57] Pourmousavi, S. A. (2018). Evaluation of the battery operation in ramp-rate control mode within a PV plant: A case study. *Solar Energy*, *166*, 242–254.

[58] Prabhu, C. (2020, June 09). https://www.omanobserver.om/. Retrieved Nov 23, 2021, from https://www.omanobserver.om/article/12733/Business/award-for-solar-diesel-hybrid-power-projects-in-oman-by-q3

[59] Prabhu, C. (2021, June 28). *PDO/EDO plan new 100MW solar power project in North Oman*. (Zawaya by Refinitiv) Retrieved Nov 23, 2021, from https://www.zawya.com/mena/en/business/story/PDOEDO_plan_new_100MW_solar_power_project_in_North_Oman-SNG_220356336/

APPENDIX 1.A: DFIG MODEL SPECIFICATION

The DFIG's specs are presented in Table 1.A.1. As seen in Table 1.A.2. The coefficients are expressed in terms of the pitch angle and tip speed ratio, as described in the literature review chapter.

TABLE 1.A.1
DFIG Specifications

Specification	Unit	Value/Type
Rated apparent power	MVA	2.2
Maximum power	MW	2.0
Rated stator voltage (L-L RMS)	V	0.69
Rated frequency	Hz	60
Rotor/stator turns ratio	-	2.6377
Inertia constant	MWs/MVA	1.5
Fraction damping	p.u	0
Rated wind speed	m/s	12
Cut in wind speed	m/s	6

TABLE 1.A.2
DFIG Model Coefficient Values

Name	Description	Value
n_1	$Cp(Imda, Beta) = (n_1 - n_2*Beta)$	0.47
n_2	$Sin(1.57*(Imda-3y)/(n_3 - n4*Beta))$	0.0167
n_3	$-(Imda - 3y)* n_5* Beta + n_6/(1 + Imda)$	7.5
n_4	*Constant par.*	0.15
n_5	$Y = 1 - exp(-Imda/3)$	0.00184
n_6	Constant par.	0.01

Table 1.A.1 contains the DFIG model's performance coefficients. Figures 1.A.1 and 1.A.2 illustrate the wind turbine's features and the effect of pitch angle incrementation on these coefficients.

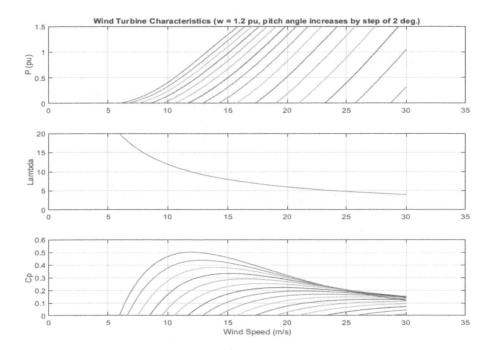

FIGURE 1.A.1   DFIG characteristics curve while incrementing pitch angle.

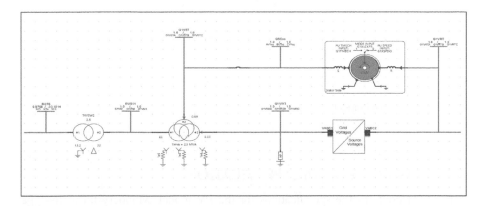

FIGURE 1.A.2   DFIG Model.

APPENDIX 1.B: DIESEL GENERATOR MODEL SPECIFICATION

Table 1.B.1 details the specification for the diesel generator model shown in Figure 1.B.1 in the RTDS.

TABLE 1.B.1
Diesel Generator Model Specifications

Specification	Unit	Value/Type
Rated apparent power	MVA	5.5
Rated voltage	kV	4
Base angular frequency	Hz	60
Inertia constant H	MWs/MVA	3.03
Synchronizing mechanical damping D	p.u	2
Machin zero sequence resistance	p.u	0.002
Machin zero sequence reactance	p.u	0.130

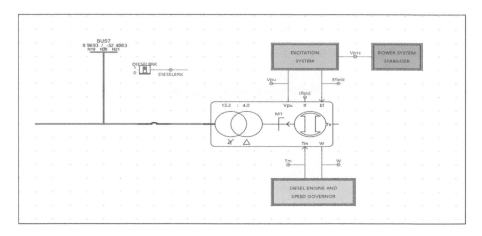

FIGURE 1.B.1   DG model.

APPENDIX 1.C: PV MODEL SPECIFICATION

Table 1.C.1 contains the specifications for the PV model. Monocrystalline/N-type cells are utilized in the PV model. The characteristic curve of the PV module's I–V conversion is shown in Figures 1.C.1 and 1.C.2. The PV model is tested under standard test settings of 1000 W/m^2 irradiance and 25°C.

TABLE 1.C.1
PV Module Specifications

Specification	Unit	Value/Type
Full maximum power	MW	1.74
MPP voltage (V_{mpp})	V	41
Open circuit Voltage (V_{oc})	V	49.1
MPP current (I_{mpp})	A	9.4
Short circuit current (I_{sc})	A	10.11
Number of modules in series × per string	No.	115×36
Number of modules in parallel × per string	No.	285×1

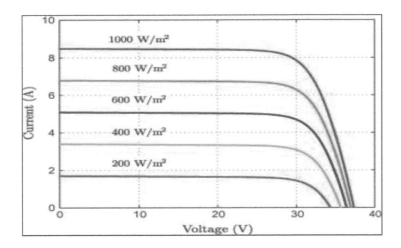

FIGURE 1.C.1   I–V characteristic curve.

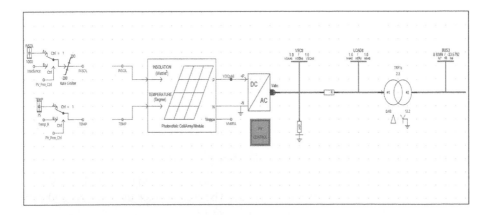

FIGURE 1.C.2   PV model.

APPENDIX 1.D: BATTERY STORAGE SYSTEM MODEL SPECIFICATION

Table 1.D.1 details the specification of the battery storage system model used in the tested MG model Shown in Figure 1.D.1.

TABLE 1.D.1
Battery Storage System Model Specifications

Specification	Unit	Value/Type
Number of cells in series in stack N_s	No.	250
Number of stack in parallel N_p	No.	250
Battery type	-	Min/Ricon-Mora
Capacity of single cell	No.	0.85
Initial stage of charging in a single cell	%	0.85

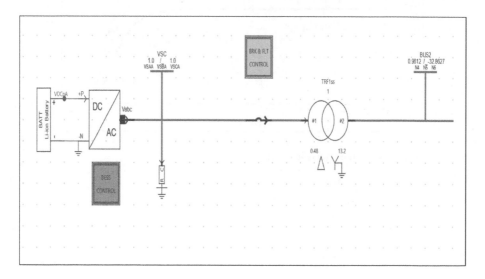

FIGURE 1.D.1 BESS model.

APPENDIX 1.E: GRID MODEL SPECIFICATION

The grid infeed voltage for the system is 138 kV and is stepped down to 13.2 kV. The grid network operates at a 60-Hz frequency. The transformer used in the grid step down has the following specifications, see Table 1.E.1.

TABLE 1.E.1
Grid Step Down Transformer Specifications

Specification	Unit	Value/Type
Based frequency	Hz	60
Transformer rating power	MVA	25
Leakage inductance of transformer	p.u	0.01
No. load losses	p.u	0
Winding 1 connection	-	Delta
Winding 1 base primary voltage (L-L RMS)	kV	138
Winding 2 connection	-	Y
Winding 2 base primary voltage (L-L RMS)	kV	13.2

1.E.1 LOAD MODEL SPECIFICATION

The MG's load is dispersed over the network. There are a total of seven dynamic loads with varying demands. The load is rated at 13.2 kV line-to-line, and all work is performed at a 60-Hz frequency. The load is controlled and modifiable in accordance with the model's test criteria.

2 Renewable Microgrid Modeling, Simulation, and Results Analysis

Muntaser Al Hasani, Amer Al-Hinai, Hassan Haes Alhelou, Ahmed Al Maashri, Hassan Yousef

CONTENTS

DOI: 10.1201/9781003307433-2

2.1 BACKGROUND

This chapter provides an overview of the process that was used to carry out the numerous research projects, as well as the foundation on which the data were gathered, the results were acquired, and the findings were analyzed. The assumptions that have been made are also highlighted.

The doubly-fed induction generator (DFIG), photovoltaics (PV), and others all have a substantial impact on the system. Because inertia-free distribution generation is increasingly being used, the synchronous generator's role is being reduced, which has an effect on system inertia. Grid-connected/grid-interconnected microgrids don't have this problem because the utility grid controls the microgrid's frequency and voltage. An isolated microgrid system, on the other hand, relies on its inertia for stability, which means that changes in system operation can lead to instability issues. When power imbalances occur in isolated microgrid systems, there is a risk of frequency instability. To solve the stability issue, a fast frequency response is needed; otherwise, the islanded microgrid's operation and control could run into serious issues.

The approach described in the following list is used to develop the controllers that are necessary for the fast frequency response and support the functioning of an islanded microgrid.

1. DFIG and PV controller set to maximum power tracking.
2. Diesel Generator (DG) and battery energy storage systems (BESS) operate in load frequency control (LFC).
3. Coordinating the DG and BESS PID controllers by tunning the controllable gain using the optimization algorithm.
4. Implementing the LFC using Real Time (RT) data from wind speed and irradiance.

Details of the followed methodology are described in the following sections.

2.1.1 DFIG MAXIMUM POWER TRACKING

Detailed advancements in microgrid controller modeling of DFIG and PV will be presented in this section. The microgrid model's resources are all connected via the controller schema.

There are two types of local controllers for wind energy: the first is called the rotor side converter (RSC), and the second is called the grid side converter (GSC). The DFIG is comprised of an induction generator and a back-to-back pulse-width modulation (PWM) converter, which includes both a rotor side and a grid side converter. When the network voltage is unbalanced, it has an impact on not just the induction generator but also the rotor's PWM converter. The RSC is normally responsible for controlling the power flow delivered to the system, whereas the GSC is responsible for managing the direct current (DC)-link bus voltage and may also have an impact on the power factor.

To reduce power and torque pulsations produced by network voltage imbalance, it is necessary to modify the control structure of the RSC's power supply. DFIG control structures can be modified to contain procedures for controlling both positive and negative sequences, in addition to the conventional control structure. Further reductions in power and torque oscillations can be achieved by incorporating a resonant harmonic controller (R) into the RSC's control loop in parallel with the proportional-integral (PI) controller. The apparent power of a stator may be defined in terms of positive and negative sequence components in an extremely unbalanced network. This is because the DFIG per-phase equivalent circuit must be examined in both the positive and negative sequence *dq*-reference frames as a result of these components.

2.1.1.1 Rotor Side Converter Control

A synchronous *dq*-axis rotating frame controls the DFIG rotor. There are a variety of ways to apply directional rotating frames. First, vector control was based on the rotational and magnetizing fluxes of the rotor [1]. However, one of the most common implementation options in contemporary DFIG is to align the *d*-axis with the location of the stator flux vectors. This is called stator-flux orientation vector control (SFO). Stator voltage orientation (SVO) is another common method of orientation [2]. Using the SFO technique, it is possible to independently modify electrical torque and rotor excitation current. As a result, both active and reactive power can be precisely controlled.

The major role of the RSC is to produce rotor excitation. The rotor currents need to be adjusted such that the rotor flux is appropriately oriented in relation to the stator flux. As a result, the DFIG's speed and torque are controlled by the in-board PWM converter.

The control structure of the RSC is depicted schematically in Figure 2.1.

The equations governing the relationship of rotor voltage (v_d and v_q) and flux ($\varnothing_d$ and $\varnothing_q$) components in *dq*-axis and under the SFO are shown as follows:

$$v_{rd} = R_r\, i_{rd} + \sigma L_r \frac{di_{rd}}{dt} - \mathcal{W}_{slip}\sigma L_r i_{rq} \tag{2.1}$$

$$v_{rq} = R_r\, i_{rq} + \sigma L_r \frac{di_{rq}}{dt} + \mathcal{W}_{slip}(L_\circ i_{ms} + \sigma L_r i_{rd}) \tag{2.2}$$

$$\varnothing_{rd} = L_\circ i_{ms} + \sigma L_r i_{rd} \tag{2.3}$$

$$\varnothing_{rq} = \sigma L_r i_{rq} \tag{2.4}$$

where $\sigma = 1 - \frac{L_m^2}{L_s L_r}$ and $L_\circ = \frac{L_m^2}{L_s}$, L_r, L_s, and L_m self-inductance of the rotor and stator, and magnetizing inductance, respectively.

As shown in Figure 2.1 and Figure 2.2, the RSC is made up of two parallel sets of PI controllers, resulting in two control loops. Speed-control loops and torque-control

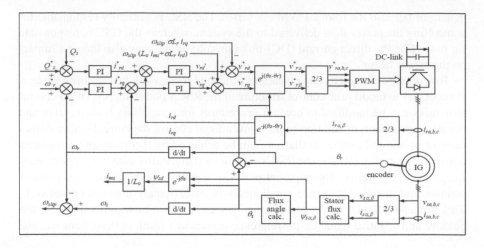

FIGURE 2.1 Control system structure of the RSC [3].

loops can both be used as the outer loop of the reference q-axis rotational current (i_{rq}^{*}). However, it is sometimes impossible to correctly measure the machine toque for torque control. As a result, when torque is applied, the system becomes an open loop. The active power output is connected indirectly to the outside control loop via the active power output (P). It is also necessary to deploy an additional PI controller to generate the d-axis rotor current reference component i_{rd}^{*}, which is used to regulate the required reactive power Q.

Stator voltage and current and rotor current are required by the controller. Final control is performed by altering output voltage of voltage source inverter (VSI)-induced rotor voltage. These V_{rd} and V_{rq} signals are sent to the appropriate PI controllers via i_{rd} and i_{rq} errors.

$$v_{rd}' = R_r\, i_{rd} + \sigma L_r \frac{di_{rd}}{dt} \tag{2.5}$$

$$v_{rq}' = R_r\, i_{rq} + \sigma L_r \frac{di_{rq}}{dt} \tag{2.6}$$

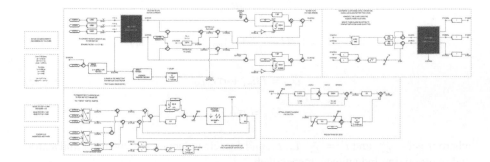

FIGURE 2.2 Controller structure of the RSC in a real-time digital simulator (RTDS).

2.1.1.2   Grid Side Converter Control

Energy flows into the grid are controlled by the GSC, which controls the flow of both active and reactive energy. Maintaining a constant voltage regardless of rotor power and direction is its key objective.

An alignment of the rotating reference frame with the stator voltage vector provides independent active and reactive power management for the GSC. The d-axis current regulates the DC-link voltage and the q-axis current regulates the reactive power of the PWM converter. Figure 2.3 and Figure 2.4 illustrate the GSC controller structure schematically.

To enable dissociated control of real and reactive power, the v_{cd} and v_{cq} equations are amended with the following compensating terms:

$$v_{cd} = R\ i_{cd} + L_{choke}\frac{di_{cd}}{dt} - \mathcal{W}_e L_{choke} i_{cd} + v_{cd1} \tag{2.7}$$

$$Cv_{cq} = R\ i_{cq} + L_{choke}\frac{di_{cq}}{dt} + \mathcal{W}_e L_{choke} i_{cq} + v_{cq1} \tag{2.8}$$

where is θ_e the grid voltage angular position and

$$v'_{cd} = R\ i_{cd} + L_{choke}\frac{di_{cd}}{dt} \tag{2.9}$$

$$v'_{cq} = R\ i_{cd} + L_{choke}\frac{di_{cd}}{dt} \tag{2.10}$$

2.1.2   PV Maximum Power Tracking

The PV is structured identically to the local control. Sources are controlled by an outer loop in grid-connected operation mode and an inner loop in current control

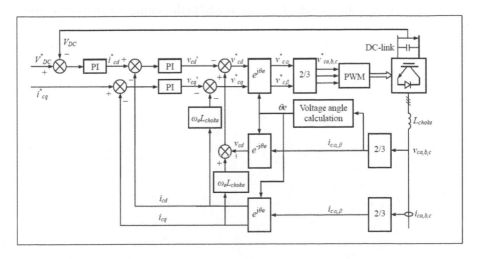

FIGURE 2.3   Control structure of the GSC [3].

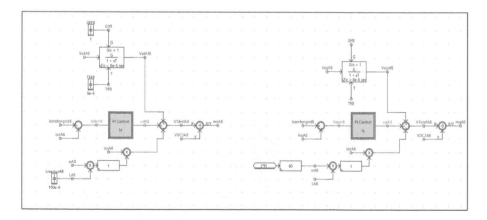

FIGURE 2.4 Controller structure of the GSC in RTDS.

mode. The inner loop control functions on a current-limited basis, whereas the outer loop control operates on a constant voltage basis. The voltage on the DC bus is controlled by the outer loop controller. The DC bus voltage is compared with a defined value, and then a PI regulator, and other techniques are used to achieve a particular current value for inner loop control. The inner loop control ensures that the PV current does not exceed the maximum permitted. The outer loop current is transferred through a controller. Whenever the current is greater than the maximum current, the output is adjusted to the maximum current; otherwise, the output is left unchanged.

Figures 2.5 and 2.6 show the whole structure of the PV inner and outer loop control systems.

2.1.3 Developed Load Frequency Control in the Microgrid System

This section explains the control designed for the BESS and DG units to achieve LFC requirement in the microgrid system, as well as the tuning of controllable gains.

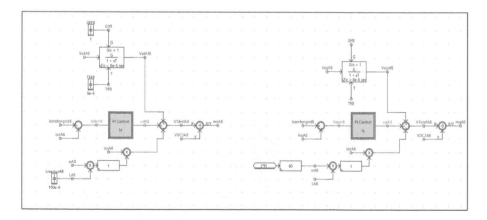

FIGURE 2.5 Inner loop *dq* current control.

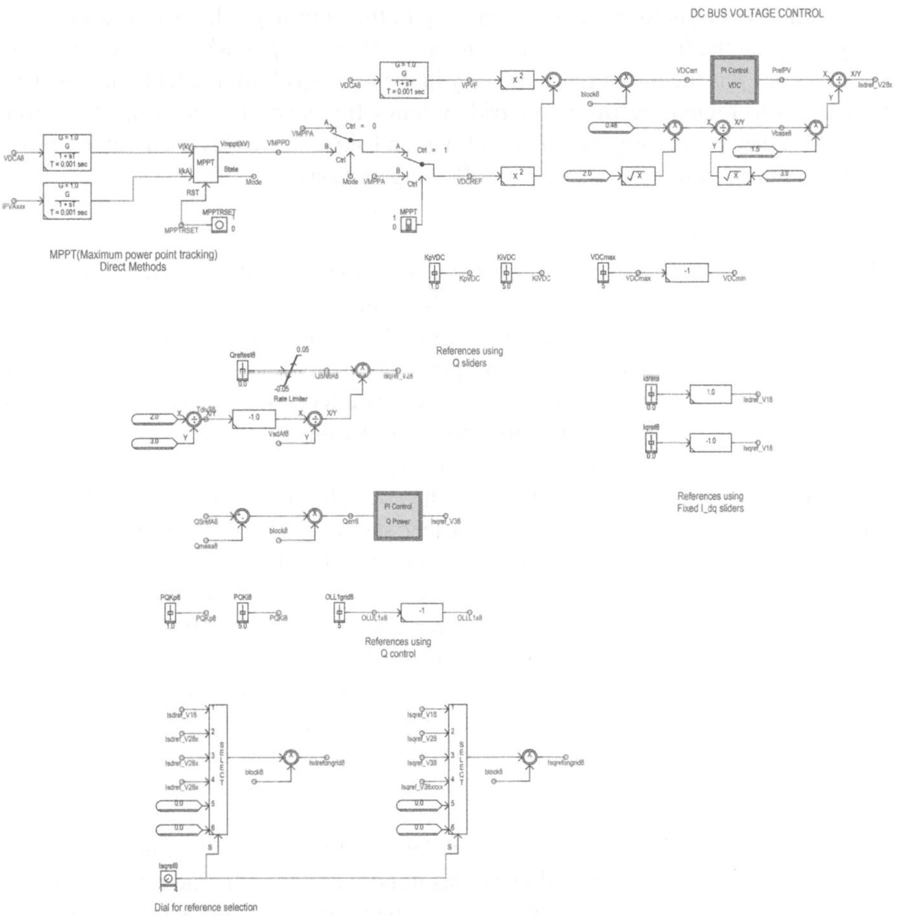

FIGURE 2.6 Outer loop grid-connected maximum power point tracking (MPPT).

The control objective is to minimize the frequency difference between the actual frequency and the nominal frequency given in Equation 2.11:

$$\Delta f = f - f^s \tag{2.11}$$

The objective function (performance index, J) is defined as follows:

$$minimize\ J = \int_0^{t_f} |\Delta f|^2 dt \tag{2.12}$$

The main goal is to reduce the frequency deviation (J) in both regular operations (island mode) and contingency scenarios, where one of the system's generating units is continuously changing due to wind speed change or solar irradiance fluctuation. As a consequence, three control loops are proposed: one on each of the two diesel engines and one on the battery storage unit to limit the frequency deviations (i.e., to minimize the performance index J).

The LFC's goal is to drive the frequency to the nominal value while simultaneously reducing the frequency deviation to zero ($\Delta f = f - f^s$), where f^s is the nominal value and f is the measured frequency in the microgrid. In line with that, the DG contributes to regulating the microgrid system's frequency by ensuring the power system balance between generation and demand sides in the MG. Therefore, three types of controls will be described in following sections.

2.1.3.1   Excitation Control System

The excitation system is a critical component of the alternating current (AC) synchronous generator because it regulates the synchronous generator's voltage and reactive power. Additionally, the speed control system regulates the prime mover generator's speed (frequency) and active power. Both are the generator set's primary control mechanisms. Excitation control system is a feedback control system that is comprised of a synchronous generator and its excitation system. The controller regulates the excitation current in response to changes in the generator load to preserve a constant terminal voltage. Additionally, the controller increases the parallel functioning of the generator's static and transient stability while the generator is subjected to a maximum and lower excitation limit based on the operation necessities.

The primary component of the excitation control system is the excitation regulator. It normally detects changes in the generator voltage and then regulates the excitation power unit. The excitation power unit does not alter its output's excitation voltage until the excitation regulator modifies the control command. Figures 2.7 and 2.8 illustrate the control structure for the diesel generator's excitation and stabilizer control in the microgrid model, which reflects the local control of the unit [4].

2.1.3.2   LFC Controller for DG

The governor's function is to regulate fuel delivery to the engine cylinders to maintain constant speed under all load situations imposed on the generator by the engine. The engine speed must be kept constant to maintain the generator's frequency.

The following formula expresses the link between the generator's frequency and the engine's speed:

where f is frequency, N is the engine speed, and P is the pole number on the generator.

$$f = N * \frac{P}{120} \tag{2.13}$$

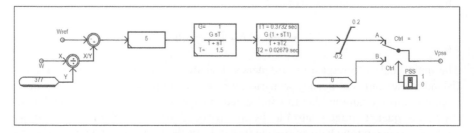

FIGURE 2.7   Power system stabilizer (PSS) control for the diesel generator.

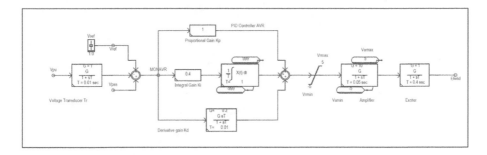

FIGURE 2.8 Excitation control structure for the diesel generator. (PID: Proportional Integral Derivative, AVR: Automatic Voltage Regulator, Vref: Reference Voltage, Vpss: output Voltage from Power System Stablizer, MONAVR: signal name, Vpu: Voltage per unit)

Figure 2.9 depicts the DG and speed governor, which uses proportional integral derivative (PID) as a speed controller.

2.1.4 BESS LOAD FREQUENCY CONTROLLER

In an isolated microgrid system, the system frequency may fluctuate over limits due to the low inertia of the installed sources. Standard EN50160/2006 [5] and [6] states that under typical operating circumstances, an islanded system's frequency must fluctuate within a range of ±2% of the system's average frequency over 10 seconds.

The P–f characteristic output is configured into a multilevel curve using the BESS's frequency control as the primary control.

The three regions (Region 1, Region 2, and Region 3) in Figure 2.10 and Figure 2.11 make up the control, and each one dictates how the BESS operates. The BESS operates according to the droop curve with a deadband zone in Region 1 for both charging and discharging. To avoid overcharging or overdischarging, the BESS can go into standby mode (inactive) in the deadband. BESS distributes electricity to Regions 2 and 3 according to the flat line when the frequency is between f_{max} and f_{min}. The synchronous generator supplies either P_{d1} (Region 2) or P_{d2} (Region 3). In Region 2, the controller maintains a frequency of f_{min}. From P_{b1} up to P_{bmax}, the BESS's power is supplied by the BESS synchronous generator. As a result, the frequency in Region 3 remains at f_{max}. From P_{b1} to P_{bmax}, the BESS power fluctuates while the generator provides P_{d2}.

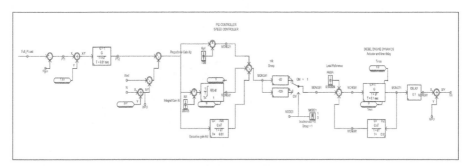

FIGURE 2.9 LFC PID controller for DG.

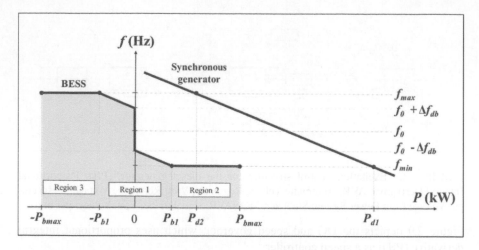

FIGURE 2.10   BESS multilevel LFC with DG droop [7].

2.1.4.1   SOC Controller

The primary issues faced by islanded microgrids are load balancing and microgrid voltage and frequency regulation, which are similar to those faced by stand-alone systems. Battery capacity and state of charge (SOC) limits must be taken into consideration in all control methods. BESS operation in each region is determined by an SOC control method in this section. The proposed technique for SOC control is depicted in Figure 2.12. According to the SOC condition, this method is separated into five control scenarios. To regulate the battery's SOC, the proposed technique may be accomplished by applying a changing droop control. The colored lines in Figure 2.13 depict the BESS with the SOC controller effect. The block diagram of the SOC controller is depicted in Figure 2.14.

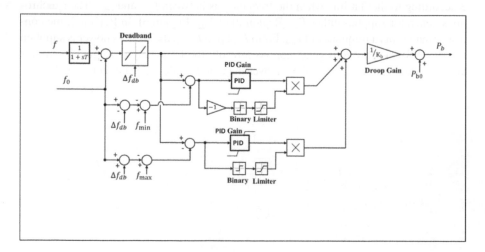

FIGURE 2.11   BESS LFC block diagram.

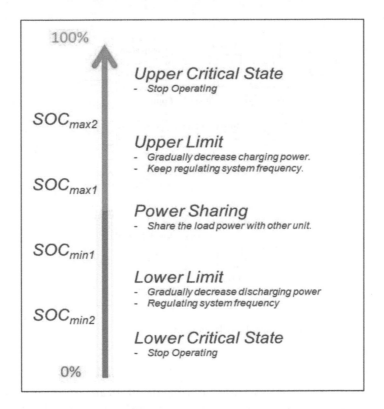

FIGURE 2.12   Proposed strategy for SOC management.

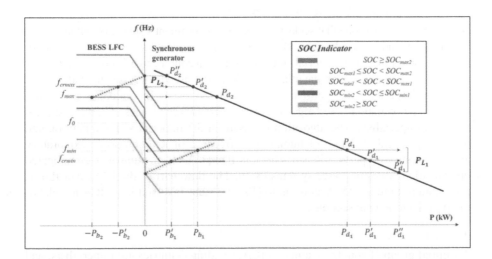

FIGURE 2.13   Schemes for SOC [7].

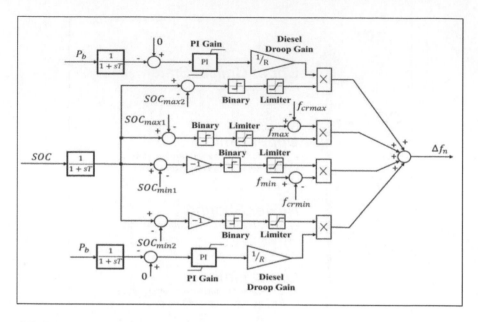

FIGURE 2.14 BESS SOC control block diagram.

2.1.5 COORDINATED LFC USING BESS AND DG CONTROLLERS

The coordinated LFC can be achieved utilizing the BESS and DG controllers. Such a process requires optimal tuning of the controllers' gains. Several optimization algorithms have been studied by researchers around the world in connection to the frequency stability in power systems. However, in this research, the gain controller is optimized using the following algorithm to improve the LFC.

2.1.5.1 Imperialist Competitive Algorithm

An imperial policy is one that seeks to broaden a government's sphere of influence and authority beyond its own borders. Direct rule or less evident tactics such as control or markets for goods or raw materials may be used by a country to exert dominance over another. Neocolonialism, on the other hand, is a term used to describe the latter.

In the imperialist competitive algorithm (ICA), every individual is a country, making it a population-based algorithm. There are two categories of countries: colonies and imperialists. The algorithm is based on the imperialistic competition and the sociopolitical evolution of humankind. Because of its appropriate exploratory and exploitative properties, ICA has been utilized to tackle a number of complicated optimization issues in power systems [8, 9]. The flowchart of the ICA algorithm is presented in Figure 2.15. According to [10], the steps involved in an ICA are elaborated in the following sections.

2.1.5.1.1 Initialization

An initial group of countries launches ICA. Leading countries are imperialists, and the rest are imperialist colonies. The imperialists divide the colonies according to

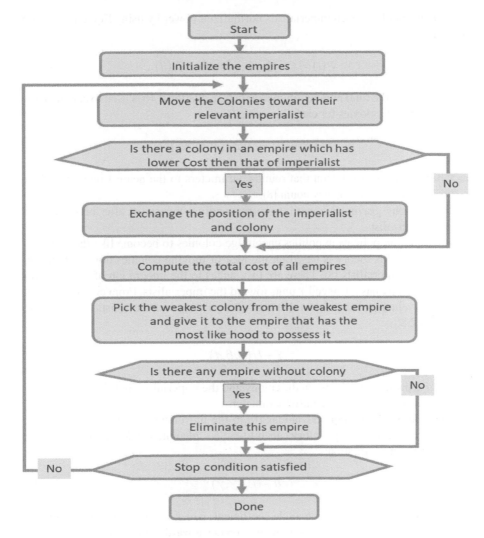

FIGURE 2.15 Flowchart of the ICA [8].

their power. The total power of an imperialist is calculated by adding the fitness of the empire and a percentage of the fitness of its colonies [8].

ICA is used to divide the colonies among imperialists as Equation 2.14 (the normal cost of an imperialist).

$$C(i) = c(i) - max(c(j)) \qquad (2.14)$$

where $c(i)$ represents the cost of ith imperialist and $C(i)$ represents the normalized cost of ith imperialist.

It then calculates each imperialist's normalized power by using Equation 2.15 on all imperialists.

$$P(i) = |C(i)\Sigma j = 1j = Nimp \ C(j)$$ (2.15)

where *Nimp* is the total number of imperialists. Equation 2.16 is used to calculate the total number of colonies for each imperialist.

$$PNcol(i) = round\left(P(i).Nc\right)$$ (2.16)

as where *round* is a function that rounds parameters to the nearest integer, and *Nc* stands for the overall country count [8].

Each colony gravitates toward its corresponding imperialist after being divided among imperialists.

Assimilation policies in politics encourage colonies to become like their imperialists; hence, this theory is based on the idea of moving colonies toward their corresponding imperialists [8]. There are two stages to the integration of ICA. By this level, the colony has advanced *Y* units toward the imperialists. From colony to imperialist, movement takes place in that way. Here, *Y* is an indeterministic variable like Equation 2.8:

$$Y \sim U\left(0, \beta.d\right)$$ (2.17)

where *d* is the distance between the colony and the imperialist, and *β* is the assimilation coefficient, which is a control parameter.

The addition of random deviations (*θ*) to the movement direction of ICA in the next step enhances its exploratory capabilities. To calculate the deviation, we need to use the following equation:

$$\theta \sim U\left(-\gamma, \gamma\right)$$ (2.18)

where *γ* stands for assimilation angle coefficient, which a control parameter [8].

Assimilation is the process of moving colonies toward their corresponding imperialists. If a colony reaches a point where it is more suited than the imperialist while traveling toward the imperialist, the imperialist and colony switch positions, and so forth.

2.1.5.1.2 *Imperialistic competition*
It is at this stage that the weakest colonies of the poorest empires are picked, and then all of the empires struggle to win control of these colonial possessions [11].

2.1.5.1.3 *Elimination of weak empires*
As a result of the imperialistic competition, weak empires will be destroyed, and their colonies will be distributed among the other empires. Most of the time, when an empire destroys all of its colonies, it is said to have dissolved and been wiped out [8].

2.1.5.1.4   Revolution

Social and political conditions in a country can shift dramatically during a revolution. It is based on this reality that each country has a certain likelihood of becoming mutated at each ICA iteration (decade). ICA's success is largely dependent on this probability, known as the "revolution rate" (r) [8].

2.1.5.1.5   Convergence

After a few iterations (i.e., decades), the most powerful empire will sustain and maintain its control over all the colonies, and all the other empires will collapse. In this case, the algorithm will stop. The diagram and pseudocode ICA can be found in [8].

2.1.6   Developing Hardware-in-the-Loop (HIL) Structure

Simulating The hardware-in-the loop (HIL) approach is the industry standard for creating and testing the most complex systems of control, protection, and surveillance. There are two major issues affecting product development now that are driving HIL's rapid growth: system complexity and delay to market.

Traditionally, control system testing was conducted directly on physical equipment in the field, on the entire system, or on a power testbed in a laboratory. Although this approach ensures testing fidelity, it may be rather costly, inefficient, and sometimes dangerous.

HIL testing is a useful solution to more traditional techniques of testing. If you utilize HIL, the physical plant is replaced by a computer model that runs in real-time digital simulator (RTDS) and is supplied with appropriate inputs and outputs (I/Os) able to connect to control systems and other equipment. It is therefore possible to perform complete closed-loop testing without having to use real-world equipment by using the HIL simulator to accurately replicate the plant and its dynamics. In addition to these qualities, HIL significantly reduces the inadequacies of current testing methods. HIL simulation has become the industry standard in a wide range of industries globally because it reduces the risk, cost, and time necessary to test complex embedded systems.

Figure 2.16 illustrates An HIL framework that models the hybrid power plant in RTDS and the central controller in Simulink Desktop Real-Time (SLDRT). The central controller performs real-time computations and transmits control signals to the several different inverters that comprise the hybrid plant model. RSCAD RunTime monitors the simulation itself (i.e., a packaged part of RSCAD). RSCAD RunTime is installed on a dedicated workstation to maximize speed and avoid overburdening the central processing unit (CPU) with other tasks. It manages the simulation and extracts findings and analysis data at a very high resolution, sampling every 50 seconds per signal and storing them in a variety of formats for transient research and monitoring. For steady-state analysis, the data obtained by Simulink at a sample rate of 1 kHz are adequate.

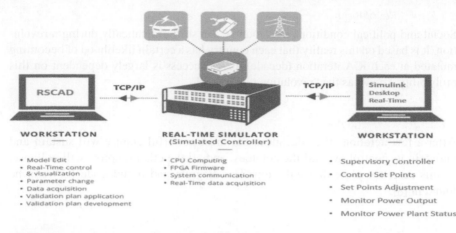

FIGURE 2.16   Hardware in the loop elements connection.

2.1.7   WIND SPEED AND SOLAR IRRADIANCE DATA

One of the most important simulation scenarios is to put the controller through its paces using real-world profile data such as wind speed and solar irradiation. This information was gathered on the ground and represents the Sultanate of Oman's data profile.

The coordinated LFC controller is the primary control that runs in a specific mode to maintain equilibrium between the entire microgrid power supply and local demand. As a consequence, the system frequency must be maintained within an acceptable range and returned to its nominal value within a relatively short period of time. Finding the optimal gains parameters for the BESS and DG using the ICA algorithm optimization approach

2.1.8   SUMMARY

This chapter provides an overview of the method that follows for designing the coordinated LFC controller. The goal is to keep the DFIG and PV controller set to maximum power tracking while DG and BESS operate in LFC, coordinating the DG and BESS PID controllers by tuning the controllable gain using the optimization algorithm and implementing the LFC using real-time data for wind speed and irradiance. All the designed details have been explored to build the highest performing LFC controller working in islanded and grid-connected operation mode.

2.2   MICROGRID MODELING, SIMULATION, AND RESULTS ANALYSIS

This section will describe the model under the test and all the simulation scenarios tested on the microgrid model. In addition, there are four main scenarios. The focus will be on increasing the load by 20% and 50%. The second scenario concentrates on the generation change due to wind speed change and solar irradiance change. The third scenario will be implementing the controller into the actual data profile for the wind speed and solar irradiance. The last scenario is to test the controller model in grid-connected mode. The results will be presented and analyzed according to the mentioned scenarios.

Various simulations were conducted to analyze the dynamic behavior of the microgrid when it was islanded, including load using the established centralized secondary LFCl approach. The effectiveness of these control mechanisms, which compose the LFC control, was evaluated using a test system in the RSCAD. The local total load of the microgrid was approximately 5.0 MW and the total generated power from the PV, DFIG, DG, and the BESS was around 5.3 MW. Some generation was modified to support frequency restoration in consideration of the observed deviation.

2.2.1 Microgrid Case Study in the RSCAD with RTDS

The full network of the microgrid modeled in RTDS using the preexisting validated and industry-approved models of DFIG, DG, PV, and BESS are connected to the grid. The model of the microgrid also consists of local loads distributed in the network. These models have been used in different studies and serve different purposes.

The capacity of each type of resource is mentioned in Table 2.1.

Detailed requirements for each energy source are provided in the chapter 1 appendices.

2.2.2 Coordinated LFC Controller Utilizing DG and BESS Compare with Uncoordinated Controller

The coordinate LFC based on PID of the DG and BESS using optimization algorithm ICA technique was compared with the uncoordinated controller. The results in Figures 2.17 and 2.18 clearly show that the coordinated controller performed better to restore the frequency to a nominal value.

2.2.3 Scenario 1: Load Change

The first scenario will test the LFC controller with a load change and different percentage increases (20% and 50%). The main operation modes for DFIG and PV are tracking the maximum power point or maintaining a constant power output. The DG and BESS will contribute more to load and generation balance to maintain the desired frequency for the microgrid in islanded mode, see Table 2.1, and Table 2.2.

TABLE 2.1
Microgrid Energy Source Capacity

Energy Source	Capacity [MW]	Voltage [kV]
DFIG	2.0	0.69
Diesel generator	5.0	4.0
PV	1.7	0.041
Battery storage system	0.2	0.041

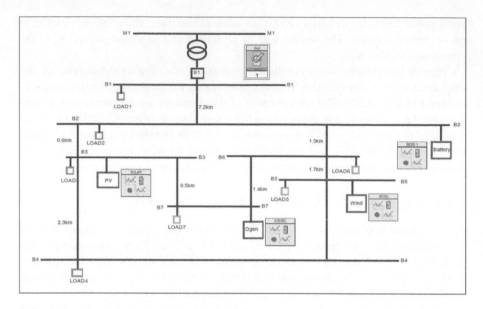

FIGURE 2.17 Full network of the microgrid model in Real Time Digital Simulator.

TABLE 2.2

Coordinate LFC vs. Uncoordinated LFC Using ICA Algorithm

Case	Settling Time Uncoordinated LFC (sec)	Settling Time Coordinated LFC (sec)	Improvements (%)
Load change 50%	7.08	2.55	64

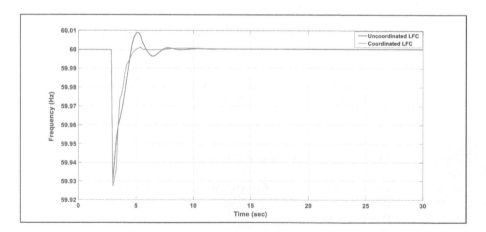

FIGURE 2.18 Frequency response with coordinated and uncoordinated LFC during load change by 50%.

TABLE 2.3

Microgrid Installed Capacity

Energy Source	Capacity [MW]
DFIG	2.0
DG	5.0
PV	1.7
BESS	0.2

2.2.3.1 Load Increase by 20%

The total installed capacity in the microgrid is around 8.7 MW plus the BESS capacity (Table 2.3).

Figure 2.19 shows a step change in the microgrid total load. The transition is characterized by a significant rise in demand of around 20%. During the operation of a power system, fluctuations in power demand are very common. The main goal of this scenario is to evaluate the behavior of the suggested LFC controller under normal operating conditions.

The frequency deviation in Figure 2.20 is caused by the load (20%) increase. However, it is essential to emphasize that the change in power demand is not solely supplied by the PV and DFIG. It is also supplied by the BESS and DG. Figure 2.21 shows that during the load change, the BESS increased from 0.2 to 2.75 MW, and it returns to nominal operation once the frequency is stabilized. The LFC for the DG acted at the same time as the load change to balance the generation shortage of the system and increased the generated power from 1.2 to 2.2 MW. The system frequency has deviated from the nominal value (60 Hz), but it is still within the acceptable range by applying the LFC controller.

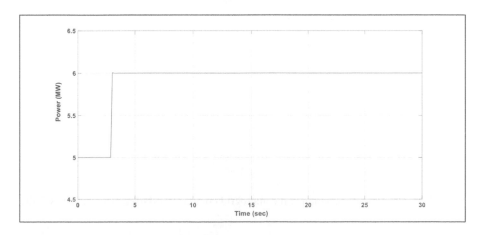

FIGURE 2.19   Step change 20% increase in power demand.

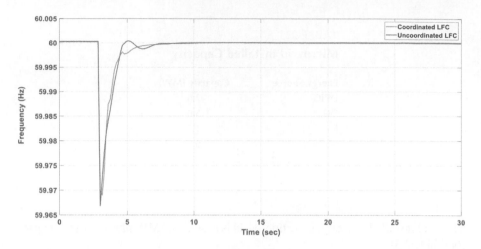

FIGURE 2.20 Frequency response with coordinated and uncoordinated LFC during load change by 20%.

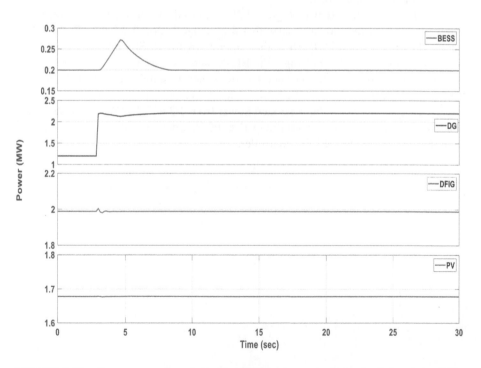

FIGURE 2.21 Power generation of the microgrid components during load change by 20%.

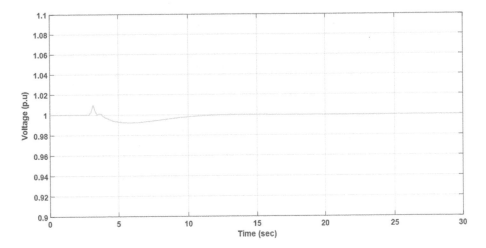

FIGURE 2.22   Voltage (p.u) of one of the busses in the microgrid network.

On the other hand, the voltage profile remains within the acceptable range of the network (the regulated range for the voltage is ±6). Figure 2.22 shows the voltage behavior during the load change. In addition, Table 2.4 ensures that all buses in the microgrid network are within the acceptable level.

2.2.3.2   Load Increase by 50%

The microgrid network is inaugurated at 60 Hz with a load requirement of 5.0 MW, which places the BESS in standby operation mode until the load demand is met. The load demand increases by 50% as the total load in the system. Local load increased from 5.0 to 7.5 MW. Figure 2.23 shows the load behavior signal during the change period. The results of the active power and system frequency are shown in Figures 2.23 and 2.24, respectively. The frequency drops as the generator attempts to maintain the balance with the load increases. The battery starts to discharge and contribute to the load demand when the frequency of the system drops. It appears from these observations that the LFC maintains the required generator output by continually adjusting the power output of the BESS.

All of the network buses are not exceeding the regulatory limit when the system voltage increases rapidly in response to a high rise in load demand.

TABLE 2.4

Voltage (p.u) of the Microgrid Buses

BUS	BUS1	BUS2	BUS3	BUS4	BUS5	BUS6	BUS7
Voltage (p.u)	0.9888	0.9889	0.9887	0.9886	0.9891	0.9885	0.9904

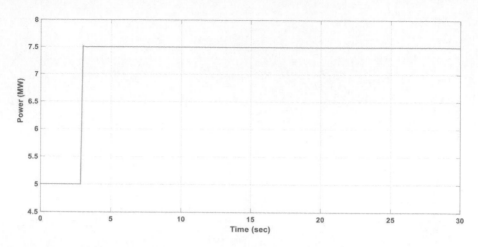

FIGURE 2.23   Step change 50% increase in power demand.

The system response based on the uncoordinated controllers of the DG and BESS is compared with the coordinated controllers. In this scenario, where the load increases by 20% and 50%, ICA coordinate controllers significantly improve both the overshot and the settling time. Tables 2.6 and 2.7 present the improvement percentage of using the coordinate method with ICA, see Figures 2.25-2.28 and Table 2.5.

2.2.4   SCENARIO 2: GENERATION CHANGE

In this scenario, the wind speed will increase from 10 to 13 m/sec, which will result in the DFIG generated power rising. The load in this scenario was kept constant at 5.0 MW. The increase in generated power in the microgrid leads to an unbalanced

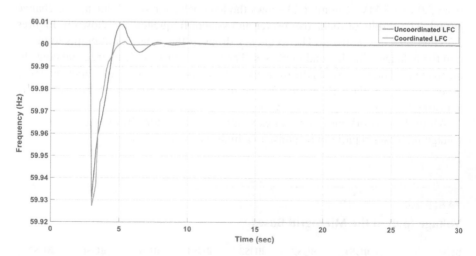

FIGURE 2.24   Frequency response with coordinated and uncoordinated LFC during load change by 50%.

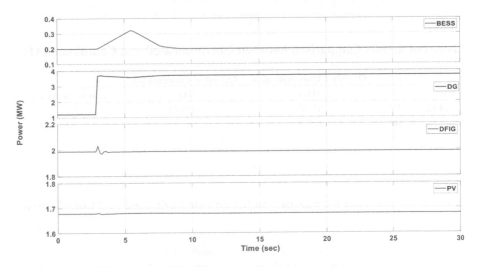

FIGURE 2.25   Power generation of the microgrid components during load change by 50%.

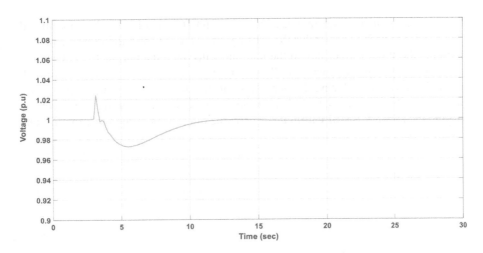

FIGURE 2.26   Voltage behavior (p.u) of Microgrid 7 buses during load change by 50%.

TABLE 2.5

Voltage (p.u) of the Microgrid Buses

BUS	BUS1	BUS2	BUS3	BUS4	BUS5	BUS6	BUS7
Voltage (p.u)	0.9773	0.9774	0.9771	0.9871	0.9776	0.9786	0.9765

TABLE 2.6

Overshot Results Comparison between Uncoordinated and Coordinated LFC

Case	Overshot Uncoordinated (Hz)	Overshot Coordinated (Hz)	Improvements (%)
Load change 20%	0.0004	0.00001	98
Load change 50%	0.009	0.001	88.89

TABLE 2.7

Settling Time Results Comparison between Uncoordinated and Coordinated LFC

Case	Settling Time Uncoordinated (sec)	Settling Time Coordinated (sec)	Improvements (%)
Load change 20%	7.05	5.25	25.5
Load change 50%	7.08	2.55	64

system. The frequency was increased to reach up to 60.0114 Hz for the coordinated LFC, and it took 8.7 seconds for the settling time. The DG and BESS participate in bringing the system frequency to the nominal value by reducing the total generated power in the microgrid where the PV is working in maximum tracking mode with negligible fluctuation during the generation of the DFIG change. The output power behavior of the DG, BESS, and PV is presented in Figure 2.29. Similarly, the voltage around the network is still within the acceptable range as presented in Table 2.8 and Figure 2.30.

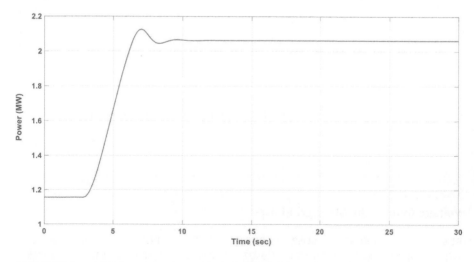

FIGURE 2.27 DFIG output power during the wind speed change.

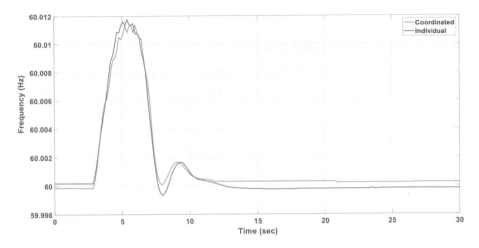

FIGURE 2.28 Frequency response with coordinated and uncoordinated LFC during generation change.

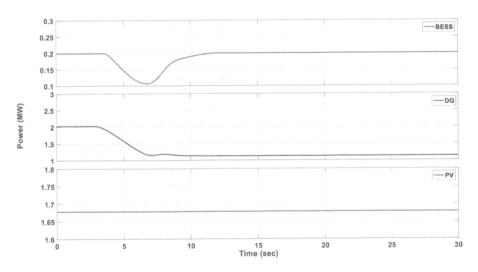

FIGURE 2.29 Power generation of the microgrid components during DFIG power change.

2.2.5 SCENARIO 3: MICROGRID LFC WITH ACTUAL DATA OF WIND SPEED AND SOLAR IRRADIANCE

In this scenario, the controller will be tested and implemented using the actual data of wind speed and solar irradiance collected from the Oman field. The load demand (5.0 MW) of the microgrid was kept constant during the full simulations.

TABLE 2.8

Voltage (p.u) of the Microgrid Buses

BUS	BUS1	BUS2	BUS3	BUS4	BUS5	BUS6	BUS7
Voltage (p.u)	0.9901	0.9902	0.990	0.990	0.9905	0.9898	0.9917

2.2.5.1   Wind Speed Profile Change

The solar irradiance and load demand were kept constant during the wind speed profile to test the LFC controller's performance.

- Solar irradiance 1,000 W/m²
- Load demand 5.0 MW

According to the following figures, the DFIG's output power varies in response to changes in wind speed. The system's stability was compromised as a result of the constant fluctuation in generated power. The uncertainty caused by wind speed causes the system frequency to vary from its nominal value, which causes the DG and BESS to react to these changes by either increasing or decreasing the generated power. Figures 2.31 and 2.32, It shows how the DG and BESS respond to all the frequency changes and try to keep the frequency within the acceptable range to stabilize the microgrid operation.

2.2.5.2   Solar Irradiance Profile Change

The wind speed and load demand were kept constant during the solar irradiance profile to test the LFC controller's performance.

- Wind speed 12 m/sec
- Load demand 5.0 MW

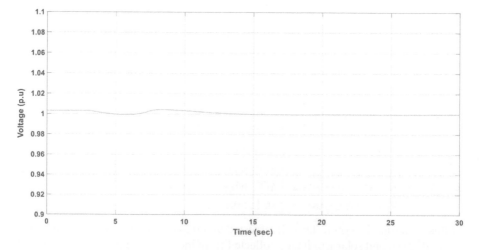

FIGURE 2.30   Voltage behavior (p.u) of Microgrid 7 buses during DFIG power change.

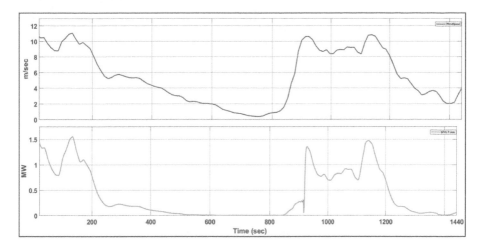

FIGURE 2.31 Wind speed profile vs. DFIG output power.

The maximum PV-generated power during the daytime is approximately 1.0 MW, whereas the power generated during the other periods due to a lack of sunlight is zero. The variation in solar irradiance affects the total generation and causes a fluctuation in the frequency of the power system. From Figures 2.33-2.34 it is clear that the BESS contributes to the microgrid by discharging or charging power to it in response to frequency deviations, in a manner similar to the DG.

2.2.5.3 Solar Irradiance and Wind Speed Change in Same Time

Combining the two profiles (wind speed profile with solar irradiance profile) is considered one of the worst scenarios to test the LFC controller's performance. The frequency deviation will be higher, and the response of the DG and BESS is clearly identified.

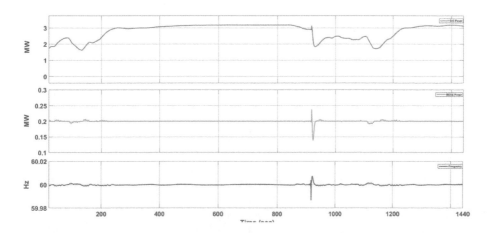

FIGURE 2.32 DG power, BESS power, and frequency response for the wind speed profile.

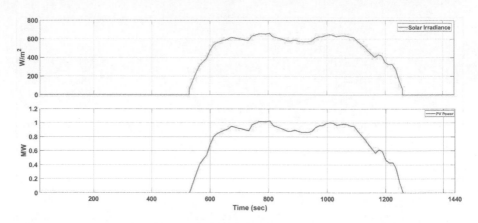

FIGURE 2.33   Solar irradiance profile vs. PV output power.

The LFC reacts to all the continuous changes in the microgrid due to the wind speed change and solar irradiance change. With all the different things going on in the system, the LFC makes sure the frequency is at a level that can be used to run the network. However, the PV and DFIG sources are working at maximum tracking power and are not contributing to the frequency deviations.

2.2.6   Scenario 4: Grid-Connected Mode

In this scenario of grid-connected operation mode, the controller works to reduce the amount of power imported from the grid when the load or the generation fluctuates.

2.2.6.1   Grid-Connected Mode with Load Change

To put this method to the test, a 50% increase in the total local load demand for the microgrid will be implemented.

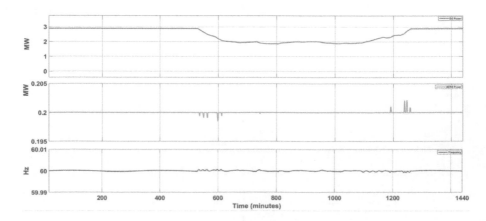

FIGURE 2.34   DG power, BESS power, and frequency response for the solar irradiance profile.

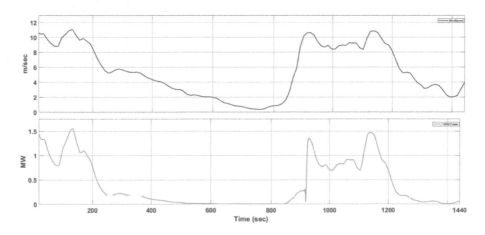

FIGURE 2.35 Wind speed profile vs. DFIG output power during comping all profiles.

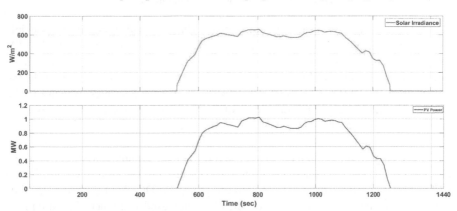

FIGURE 2.36 Solar irradiance profile vs. PV output power during comping all profiles.

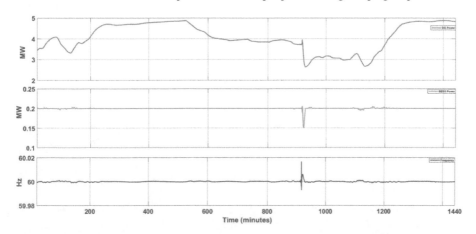

FIGURE 2.37 DG output power, BESS output power, and system frequency response, during combining two profiles.

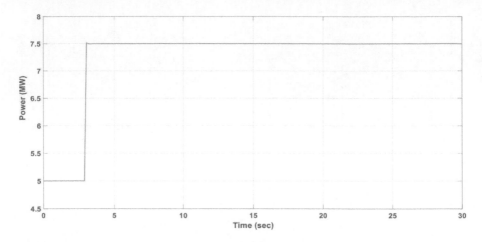

FIGURE 2.38   Load increased by 50% in grid-connected mode.

The grid-connected operation mode means that the grid is mainly responsible for maintaining system frequency. The renewable energy resources (RESs) (PV and DFIG) are working in maximum power point tracking (MPPT) mode with a total generated power of around 3.65 MW. The DG is in droop mode while the BESS started working in charging mode because there is a surplus capacity in the microgrid. Thus, the starting operation of the system is in the mode of exporting power to the grid by around 0.3 MW.

Once the load reaches 7.5 MW, the grid will step in to provide the additional load necessary to maintain a balance between the load demand and the generation. However, to reduce the amount of imported power from the grid, the coordinated LFC controller raises the generated power from the DG from 1.86 to 2.2 MW. Additionally, the generated power of BESS is increased to reduce the amount of

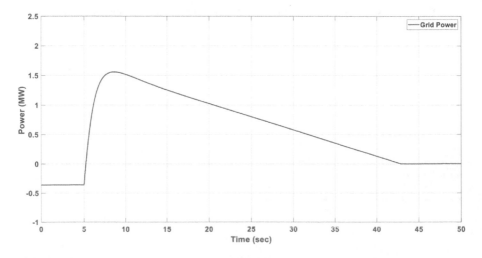

FIGURE 2.39   Import and export power to the grid.

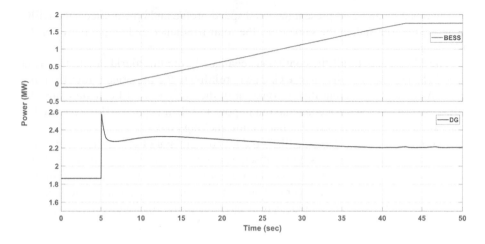

FIGURE 2.40   Generated power from the BESS and DG.

imported power into the grid to a bare minimum. Figure 2.40 shows the BESS and
DG reaction to the load increased by 50%. Figure 2.41 shows the RES generated
power in the microgrid where there is no reaction except low variation during the
same time of load change, then return back to the nominal mode.

2.2.6.2   Grid-Connected Mode with Wind and Solar Variation

Combining the two profiles (wind speed profile and irradiance profile) is considered
to be one of the worst scenarios for testing the performance of the LFC controller.
The grid regulates the frequency and voltage of the system, and the reaction of the
DG and BESS is to limit the import of power from the grid during load changes and
when there is a generating shortage.

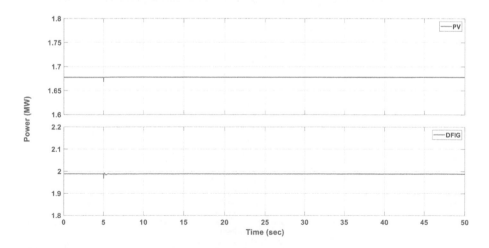

FIGURE 2.41   Generated power from the PV and DFIG.

Figures 2.42 and 2.43 present the wind speed profile and how it affects the DFIG generated power and, at the same time, the solar irradiance and output power of the PV system.

The LFC reacts to all the continuous changes in the microgrid due to the wind speed change and solar irradiance change to minimize the imported power from the grid. With all the different things going on in the system, the LFC makes sure the BESS and DG will generate the maximum power of their capacity. However, the PV and DFIG sources are working at maximum tracking power.

According to Figure 2.44, the BESS reduces the amount of power that is imported into the microgrid by increasing the discharging power to a maximum of 2.0 MW, which is the maximum generated power of the BESS. When the DG is operating in droop mode, it increases the generated power to a certain extent, but the DG's output in grid-connected operation mode is restricted due to the cost of diesel.

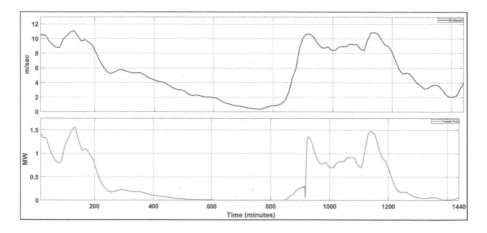

FIGURE 2.42   Wind speed profile vs. DFIG output power during grid-connected mode.

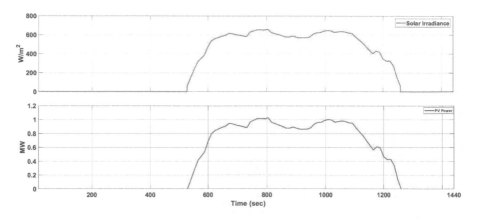

FIGURE 2.43   Solar irradiance profile vs. PV output power during grid-connected mode.

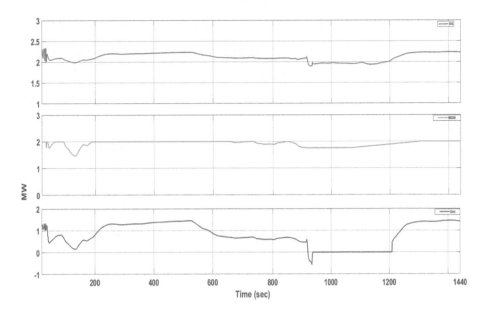

FIGURE 2.44 DG output power, BESS output power, and grid power, in grid-connected mode.

2.3 SUMMARY

This chapter presented four scenarios used to test and validate the designed strategic controller performance. Initially, the coordinate LFC was based on the PID of the DG and BESS using the optimization algorithm ICA technique and compared with the uncoordinated controller. The coordinated controller performs better (in terms of settling time and overshot) to restore the frequency to the nominal value. Second, the first and second scenarios tested the coordinated LFC with sudden changes in load (an increase of 20% and 50%) and sudden changes in generation (generated power by PV and DFIG) to check the performance of the controller with uncertainty in islanded operation mode. The third scenario tested the developed controller with real data to represent the wind speed and solar irradiance of Oman as a profile in islanded operation mode. Last, the controller is designed to minimize the imported power from the grid during the grid-connected mode, and for this reason, the fourth scenario was performed so the microgrid could be connected to the grid by increasing the load by 50% as the first sub-scenario under the grid-connected scenario. The second sub-scenario applied the wind speed profile and solar irradiance profile in grid-connected mode.

The results are presented in terms of the generated power of each unit source, the frequency of the system, and the total voltage across each system bus.

REFERENCES

[1] Regulation, Authority for Public Services (n.d.). *PRODUCT APPLICATION FOR SAHIM 2*. (APSR) Retrieved December 2, 2021, from https://www.apsr.om/en/product-application-for-sahim2

[2] RTDS (n.d.). *RTDS*. Retrieved Jan 21, 2019, from www.rtds.com/applications/microgrid-simulation-testing/

[3] RTDS Winnipeg, MB, Canada (2014). *Real-Time Simulation*. Retrieved from http://www.rtds.com.

[4] S. M. Kaviri, M. Pahlevani, P. Jain, and A. Bakhshai (2017). A review of AC microgrid control methods. *IEEE 8th International Symposium on Power Electronics for Distributed Generation Systems (PEDG)*. Florianopolis, Brazil.

[5] S. Müller, M. Deicke, and R. W. De Doncker. (2002). Doubly fed induction generator systems for wind turbines. *IEEE Ind. Appl. Mag.*, *8*, 26–33.

[6] Sandro Sitompul, Yuki Hanawa, Verapatra Bupphaves, and Goro Fujita (8 September 2020). State of charge control integrated with load frequency control for BESS in islanded microgrid. *Energies 2020*, *13*(18), 4657; https://doi.org/10.3390/en13184657

[7] Shi H., Zhuo F., and Yi H. (2015). A novel real-time voltage and frequency compensation strategy for photovoltaic-based microgrid. *IEEE Transactions on Industrial Electronics*, *62*(6), 3545–3556.

[8] Srivastava A., Ravikumar K. G., and Zweigle G. (2011). Wide-area monitoring and control using the real time digital simulator and a synchrophasor vector processor. *European Transactions on Electrical Power*, *21*(4), 1521–1530.

[9] Suganthan, S. D. (2011, Feb). Differential evolution: A survey of the state-of-the-art. *IEEE Transactions on Evolutionary Computation*, *15*(1), 4–31. doi:10.1109/TEVC.2010.2059031.

[10] Atif, A., and Khalid, M. (2020). Saviztky–Golay filtering for solar power smoothing and ramp rate reduction based on controlled battery energy storage. *IEEE Access*, *8*, 33806–33817.

3 Intelligent Control Design for PV Grid-Connected Inverter

Myada Shadoul, Hassan Yousef,
Rashid Al-Abri, Amer Al-Hinai

CONTENTS

DOI: 10.1201/9781003307433-3

3.1 INTRODUCTION TO PV GRID-CONNECTED INVERTER CONTROL

To avoid the costly expense of developing new or expanded facilities, photovoltaic (PV) solar panels, wind turbines, and other renewable energy sources (RESs) are being incorporated into existing power systems [1]. Direct current-alternating current (DC-AC) inverters are the final step in the PV system integration process. To maintain network stability and provide acceptable dynamic voltage and frequency performance, special attention for inverter topologies and controls is necessary [2]. Previous research has looked at several controllers for grid-connected inverters in islanded and grid-connected operation modes [3, 4]. Based on the performance and operating parameters of the electrical grid, the inverter controllers may be classified as linear, nonlinear, adaptive, predictive, and intelligent [5].

Classical controllers have all been described as linear controllers for microgrid inverters [6–8]. In [9–13], grid-connected inverter systems (GCIS) nonlinear controllers were presented. The H∞ robust control approach is used to propose a current-control scheme for voltage source inverters (VSIs) in [14]. In addition, [15–18] reported on adaptive and model predictive controllers for GCIS. When compared with the performance of linear controllers, the proposed nonlinear controllers outperformed them in all of the studies. The fundamental disadvantage of nonlinear control techniques is the fact that they are dependent on the availability of the mathematical model of the system and the parameters of the system.

For nonlinear systems control, intelligent control systems are introduced including fuzzy logic controllers (FLCs) and neural network controllers. Intelligent controllers have the benefit of not relying on the system mathematical model and being able to manage a wide range of nonlinear and unpredictable systems. Type 1 and type 2 FLCs are extensively employed in a variety of applications and have shown to be effective in dealing with increasing degrees of uncertainty [19–24]. Intelligent control for different applications including robotics, Internet bandwidth control, and DC-DC converters is among the FLC uses. In the health area, type 2 fuzzy logic has also shown to be effective in clinical diagnosis, differential diagnosis, and nursing evaluation [25].

Different FLCs for GCIS were introduced in [26–28]. For three-phase GCIS real-time FLC, testing to manage voltage and current was suggested in [27]. FLC's capability to produce high-quality PV electricity to retain a unity power factor was proved by these findings. In [28], a grid side inverter system control employing a basic FLC was presented, which works well for grid coupled variable speed wind turbines. Type 2 FLC (T2FLC) was used to regulate a DC-DC buck converter in another study [29]. In [30], an interval T2FLC approach was presented for PV systems.

Furthermore, in [31] T2FLC is used as a maximum power point tracking (MPPT) to deal with the rules' ambiguities through extreme weather conditions. The suggested MPPT demonstrated a quicker transient reaction and a stable steady state. Interval T2FLC (IT2FLC) was also utilized as the MPPT algorithm in [32], which was created for single-phase grid-linked PV systems. The suggested IT2FLC-based MPPT controller exhibits a quick transient response, according to simulation findings. For PV application inverters, an FLC was described in [33] that examined a number of issues and problems and provided advice for creating competent and effective inverter control systems.

Furthermore, [34, 35] reported controlling the active and reactive power of three-phase grid-connected PV systems using a fuzzy neural network controller during grid disturbances.

Furthermore, adaptive fuzzy control (AFC) approaches for controlling uncertain nonlinear systems have been developed to alleviate the shortcomings of traditional controllers for uncertain nonlinear systems [36, 37]. Researchers have employed AFC approaches in a variety of applications because it deals with complicated uncertain nonlinear systems [38–42]. The applications include induction motor control [38], wind energy conversion systems control [39], and investigating permanent magnet synchronous motor control and fuzzy fault-tolerant switching systems [40, 41].

In this chapter, we'll look at how to control a grid-connected inverter with an adaptive fuzzy approximation control scheme by introducing two papers that describe the controller idea and how to apply it to a grid-connected inverter. Adaptive fuzzy approximation control of PV grid-connected inverters and adaptive interval type 2 fuzzy tracking control of PV grid-connected inverters are the first and second papers, respectively.

3.2 ADAPTIVE FUZZY APPROXIMATION CONTROL OF PV GRID-CONNECTED INVERTERS

3.2.1 INTRODUCTION

There is no published work that describes the application of AFC to the grid-connected inverter to the best of the authors' knowledge. This prompts the authors to suggest an AFC approach that takes use of the multi-input multi-output (MIMO) feedback linearization idea as well as fuzzy systems' approximation ability. Because of the alternating nature of PV and the inverter pulse-width modulation (PWM) technology, PV GCISs are extremely nonlinear and unpredictable systems. These irregularities and nonlinearities cause many challenges without a fast-acting inverter controller. Due to the approach's capability to handle complicated nonlinear control systems without the requirement for the system mathematical model, the suggested AFC is established. The controller will be designed using the fuzzy system's ability to estimate unknown GCIS parameters for various operating circumstances. The suggested AFC for GCIS has two goals: to manage the power factor and the DC voltage. For several simulation instances, the feature of the developed controller will be verified to confirm its efficacy.

A comparison of the suggested controller's performance, that of the proportional-integral (PI) controller, and that of a fuzzy neural network controller, was also undertaken to assess its efficiency. The following is the paper's main contribution:

- An adaptive fuzzy approximation control technique for GCIS is proposed.
- The suggested controller achieves excellent tracking performance under a variety of operating situations, including power factor, parameter, and modeling uncertainty.

The remainder section 3.2 of the chapter is divided as: The feedback linearization and GCIS MIMO are given in Section 3.2.2. The design of an AFC for a generic MIMO is presented in Section 3.2.3. The suggested AFC for GCIS is discussed in Section 3.2.4 based on the analysis given in Section 3.2.3. In Section 3.2.5, the simulation results are provided, and in Section 3.2.6, conclusions are reached.

3.2.2   THE GRID-CONNECTED INVERTER SYSTEM (GCIS)

3.2.2.1   GCIS Model

Figure 3.1 depicts a GCIS. A PV array, a DC-link capacitor, a three-phase VSI, and a three-phase grid make up the system. The PV system's output power is an extremely nonlinear and unpredictable system. The MPPT approach is typically used to modify the PV array output voltage to achieve the maximum possible power regardless of solar irradiation or cell temperature. Furthermore, the MPPT technique supports the release of the DC-link voltage reference instruction [43]. Many MPPT strategies for PV systems have been documented, but the perturb and observe (P&O) and incremental conductance (IC) procedures are the most often employed in practice [44].

3.2.2.2   MIMO Model of GCIS

Figure 3.1 shows a model of the GCIS that may be expressed by

$$v_a = Ri_a + L\frac{di_a}{dt} + v_{ga} \tag{3.1}$$

$$v_b = Ri_b + L\frac{di_b}{dt} + v_{gb} \tag{3.2}$$

$$v_c = Ri_c + L\frac{di_c}{dt} + v_{gc} \tag{3.3}$$

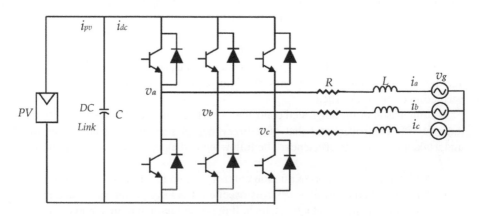

FIGURE 3.1   Three-phase grid-connected inverter.

where grid voltage and current components are v_{ga}, v_{gb}, and v_{gc} and i_a, i_b, and i_c, and inverter output voltage are is v_a, v_b, and v_c [45]. In the rotating dq frame Equations (3.1)–(3.3) are formed as

$$v_d = v_{gd} + Ri_d + L\frac{di_d}{dt} + \omega Li_q \tag{3.4}$$

$$v_q = v_{gq} + Ri_q + L\frac{di_q}{dt} + \omega Li_d \tag{3.5}$$

where dq grid voltage and current components are v_{gd}, v_{gq}, i_d, and i_q, and the dq inverter output voltage components are v_d and v_q. The connection between the DC input and AC output sides is made by [45]

$$v_{gd}i_d + v_{gq}i_q = v_{dc}i_{dc} \tag{3.6}$$

$$C\frac{dv_{dc}}{dt} = i_{pv} - i_{dc} = i_{pv} - \frac{v_{gd}i_d + v_{gq}i_q}{v_{dc}} \tag{3.7}$$

where v_{dc} is the PV output voltage, i_{pv} is the PV output current, and i_{dc} is the inverter input current.

Specifying x and u as

$$x = \begin{bmatrix} x_1 \\ x_2 \\ x_3 \end{bmatrix} = \begin{bmatrix} i_d \\ i_q \\ v_{dc} \end{bmatrix} \tag{3.8}$$

$$u = \begin{bmatrix} u_1 \\ u_2 \end{bmatrix} = \begin{bmatrix} v_d \\ v_q \end{bmatrix} \tag{3.9}$$

The GCIS state model may then be constructed as shown in Equation (3.10).

$$\dot{x} = \begin{bmatrix} -\dfrac{R}{L}x_1 + \omega x_2 - \dfrac{v_{gd}}{L} \\[2mm] -\dfrac{R}{L}x_2 - \omega x_1 - \dfrac{v_{gq}}{L} \\[2mm] \dfrac{i_{pv}}{C} - \dfrac{v_{gd}x_1 + v_{gq}x_2}{Cx_3} \end{bmatrix} + \begin{bmatrix} \dfrac{1}{L} & 0 \\[2mm] 0 & \dfrac{1}{L} \\[2mm] 0 & 0 \end{bmatrix} u \tag{3.10}$$

The control purpose is to control the grid's power factor DC input voltage v_{dc}. As a result, the system's output vector is defined as

$$y = \begin{bmatrix} y_1 \\ y_2 \end{bmatrix} = \begin{bmatrix} x_2 \\ x_3 \end{bmatrix} = \begin{bmatrix} i_q \\ v_{dc} \end{bmatrix} \tag{3.11}$$

Equations (3.10) and (3.11) can be represented as the MIMO system's general formulation as

$$\dot{x} = f(x) + g(x)u, \quad y = h(x) \tag{3.12}$$

$f(x)$ and $g(x)$ are defined by

$$f = \begin{bmatrix} f_1 \\ f_2 \\ f_3 \end{bmatrix} = \begin{bmatrix} -\dfrac{R}{L}x_1 + \omega x_2 - \dfrac{v_{gd}}{L} \\[2mm] -\dfrac{R}{L}x_2 - \omega x_1 - \dfrac{v_{gq}}{L} \\[2mm] \dfrac{i_{pv}}{C} - \dfrac{v_{gd}x_1 + v_{gq}x_2}{Cx_3} \end{bmatrix}, \quad g = \begin{bmatrix} \dfrac{1}{L} & 0 \\[2mm] 0 & \dfrac{1}{L} \\[2mm] 0 & 0 \end{bmatrix} \tag{3.13}$$

By the input-output feedback linearization technique, the MIMO model [Equations (3.12) and (3.13)] of the GCIS may be transformed to a feedback linearizable structure. To turn the nonlinear system dynamics to decoupled linear subsystems, a nonlinear control signal is constructed and applied in this technique. Following that is the feedback linearization for GCIS.

3.2.2.3   GCIS Input-Output Feedback Linearization

We used the concept of relative degree to create a feedback linearization control. Each output is successively differentiated until one input, u_1 or u_2, appears [46]. For the first output y_1 it can be demonstrated that the relative degree r_1 is $r_1 = 1$ and for the second output y_2 the relative degree r_2 is $r_2 = 2$ as

$$\dot{y}_1 = f_2 + \frac{1}{L}u_2 \tag{3.14}$$

$$\ddot{y}_2 = \dot{f}_3 = \frac{1}{C}\frac{di_{pv}}{dt} - \frac{1}{C.x_3}\left[v_{gd}\left(f_1 + \frac{1}{L}u_1\right) + v_{gq}\left(f_2 + \frac{1}{L}u_2\right)\right] + \frac{\left(v_{gd}x_1 + v_{gq}x_2\right)}{Cx_3^2}f_3 \tag{3.15}$$

Then Equations (3.14) and (3.15) may be written as

$$\begin{bmatrix} \dot{y}_1 \\ \ddot{y}_2 \end{bmatrix} = \alpha(x) + \beta(x)\begin{bmatrix} u_1 \\ u_2 \end{bmatrix} \tag{3.16}$$

where

$$
\alpha(x) = \begin{bmatrix} f_2 \\ m - \dfrac{1}{Cx_3}\left(v_{gd}f_1 + v_{gq}f_2\right) + \dfrac{\left(v_{gd}x_1 + v_{gq}x_2\right)}{Cx_3^2}f_3 \end{bmatrix} \tag{3.17}
$$

$$
\beta(x) = \begin{bmatrix} 0 & \dfrac{1}{L} \\ -\dfrac{v_{gd}}{LCx_3} & -\dfrac{v_{gq}}{LCx_3} \end{bmatrix} \tag{3.18}
$$

and $m = \frac{1}{C}\frac{dI_{pv}}{dt}$.

When Equation (3.19) is applied to Equation (3.16), The following is obtained

$$
\begin{bmatrix} u_1 \\ u_2 \end{bmatrix} = \beta^{-1}(x) \begin{bmatrix} v_1 - \alpha_1 \\ v_2 - \alpha_2 \end{bmatrix} \tag{3.19}
$$

$$
\begin{bmatrix} \dot{y}_1 \\ \ddot{y}_2 \end{bmatrix} = \begin{bmatrix} v_1 \\ v_2 \end{bmatrix} \tag{3.20}
$$

where v_1 and v_2 are a signal that may be selected in such a way that the outputs track asymptotically y_1 and y_2 to their references $y_{ref1} = i_{qref}$ and $y_{ref2} = v_{dcref}$. Defining the tracking errors $e_1 = \left(y_{ref1} - y_1\right)$ and $e_2 = \left(y_{ref2} - y_2\right)$, the signals v_1 and v_2 can be selected as

$$
v_1 = k_{01}e_1 + \dot{y}_{ref1} \tag{3.21}
$$

$$
v_2 = k_{02}e_2 + k_{12}\dot{e}_2 + \ddot{y}_{ref2} \tag{3.22}
$$

Substituting Equations (3.21) and (3.22) into Equation (3.20) yields the next tracking error dynamics:

$$
\dot{e}_1 + k_{01}e_1 = 0 \tag{3.23}
$$

$$
\ddot{e}_2 + k_{12}\dot{e}_2 + k_{02}e_2 = 0 \tag{3.24}
$$

The coefficients k_{01}, k_{02}, and k_{12} are design factors chosen such that the characteristic polynomials of Equations (3.23) and (3.24) are Hurwitz, guaranteeing that the tracking errors e_1 and e_2 converge to zero asymptotically [46].

The control rule in Equation (3.19) has a major drawback in that the precise system parameters in $\alpha(x)$ and $\beta(x)$ must be identified, and each change in the parameters has an influence on the output of the controller. The GCIS parameters may be unknown or inaccurate in practice, and ambiguity in these parameters is unavoidable. To circumvent this limitation, the fuzzy systems' universal approximation capacity is employed to estimate $\alpha(x)$ and $\beta(x)$. The suggested adaptive fuzzy controller for GCIS is provided in the next section.

3.2.3 THE PROPOSED ADAPTIVE FUZZY CONTROLLER

3.2.3.1 Adaptive Fuzzy Approximation Controller for GCIS

In this section the suggested controller is created utilizing Equation (3.12). Equation (3.16) shows the input-output feedback linearization with $r_1 = 1$ and $r_2 = 2$ could be written in the following format:

$$y^{(r)} = \alpha(x) + \beta(x)u \qquad (3.25)$$

where $y^{(r)} = \begin{bmatrix} y_1^{(r_1)} \\ y_2^{(r_2)} \end{bmatrix} = \begin{bmatrix} \dot{y}_1 \\ \ddot{y}_2 \end{bmatrix}$, $u = \begin{bmatrix} u_1 \\ u_2 \end{bmatrix}$, $\alpha(x)$ and $\beta(x)$ are the same as in Equations (3.17) and (3.18).

With a singleton fuzzifier, weighted average defuzzifier, and product inference rule a fuzzy logic system (FLS) was used to estimate the functions $\alpha(x)$ and $\beta(x)$. The concept of the fuzzy basis function (FBF) expansion $\xi(x)$ was employed to create these estimations [36] as

$$\hat{\alpha}_i(x) = \theta_i^T \xi(x) \qquad (3.26)$$

$$\hat{\beta}_{ij}(x) = \theta_{ij}^T \xi(x) \qquad (3.27)$$

where $\theta_i \in R^{M \times 1}$ and $\theta_{ij} \in R^{M \times 1}$ depict vectors of variables that may be changed and $\xi(x) \in R^{M \times 1}$ indicates the FBFs vector. The weighted-average defuzzifier was used to create the FBF [47].

$$\xi_i(x) = \frac{\prod_{i=1}^{n} x_i \, \mu_{il}(x_i)}{\sum_{l=1}^{M} \left(\prod_{i=1}^{n} \mu_{il}(x_i) \right)} \qquad (3.28)$$

where the number of states is n, and M is number of If-Then rules.

By substituting $\alpha(x)$ and $\beta(x)$ with their respective fuzzy estimations we find

$$y^{(r)} = \hat{\alpha}(x) + \hat{\beta}(x)u \qquad (3.29)$$

where $\hat{\alpha}(x) = \begin{bmatrix} \hat{\alpha}_1 \\ \hat{\alpha}_2 \end{bmatrix}$ and $\hat{\beta}(x) = \begin{bmatrix} \hat{\beta}_{11} & \hat{\beta}_{12} \\ \hat{\beta}_{21} & \hat{\beta}_{22} \end{bmatrix}$.

As a result, the AFC may be stated as

$$u = \hat{\beta}^{-1}(x)\left(y_{ref}^{(r)} + K\underline{e} - \hat{\alpha}(x) \right) \qquad (3.30)$$

where $y_{ref}^{(r)} = \begin{bmatrix} y_{ref1}^{(n)} & y_{ref2}^{(r_2)} \end{bmatrix}^T$, $K = diag[k_1 \ k_2]$, $\underline{e} = \begin{bmatrix} \underline{e}_1 & \underline{e}_2 \end{bmatrix}^T$, $k_1 = k_{01}$, $k_2 = [k_{02} \ k_{12}]$, $\underline{e}_1 = e_1$, $\underline{e}_2 = [e_2 \ \dot{e}_2]$, and $e_i = y_{refi} - y_i$, $i = 1, 2$.

3.2.3.2　Closed-Loop Stability

In this section the Lyapunov function evaluation is used to demonstrate the boundedness of the tracking error and the customizable parameters. Equation (3.30) may be rewritten in the following way:

$$\hat{\beta}(x)u = \left(y_{ref}^{(r)} - y^{(r)}\right) + K\underline{e} + y^{(r)} - \hat{\alpha}(x) \tag{3.31}$$

Equation (3.25) is substituted into Equation (3.31), and in terms of fuzzy approximation errors, the following error equation is obtained:

$$\begin{bmatrix} e_1^{(n)} \\ e_2^{(r2)} \end{bmatrix} = \left(y_{ref}^{(r)} - y^{(r)}\right) = -K\underline{e} + \left(\hat{\alpha}(x) - \alpha(x)\right) + \left(\hat{\beta}(x) - \beta(x)\right)u \tag{3.32}$$

The error equation for ith output becomes

$$e_i^{r_i} = -k_i\underline{e}_i + \Delta\alpha_i(x) + \sum_{j=1}^{p}\Delta\beta_{ij}(x)u_j \tag{3.33}$$

where $\Delta\alpha_i(x) = \hat{\alpha}_i(x) - \alpha_i(x)$ and $\Delta\beta_{ij}(x) = \hat{\beta}_{ij}(x) - \beta_{ij}(x)$ are the fuzzy estimate errors.

The error equation of the ith output in state-variable form

$$\underline{\dot{e}}_i = A_i\underline{e}_i + [\Delta\alpha_i(x) + \sum_{j=1}^{p=2}\Delta\beta_{ij}(x)u_j]b_i \tag{3.34}$$

where A_i and b_i are provided by

$$\begin{cases} A_1 = -k_{01}, \; b_1 = 1 \\ A_2 = \begin{bmatrix} 0 & 1 \\ -k_{12} & -k_{02} \end{bmatrix}, \; b_2 = \begin{bmatrix} 0 \\ 1 \end{bmatrix} \end{cases} \tag{3.35}$$

Theorem 1. *If the update rules of the parameter vectors $\theta_i \in R^{M\times1}$ and $\theta_{ij} \in R^{M\times1}$ are selected as in Equations (3.36) and (3.37), the closed-loop tracking error $\underline{e} = \begin{bmatrix} \underline{e}_1 & \underline{e}_2 \end{bmatrix}^T$ is globally eventually confined:*

$$\dot{\theta}_i = -\gamma_i\underline{e}_i^T P_i b_i \xi(x) \tag{3.36}$$

$$\dot{\theta}_{ij} = -\gamma_{ij}\underline{e}_i^T P_i b_i \xi(x)u_j \tag{3.37}$$

where γ_i and γ_{ij} denote design parameters and P_i denotes a single positive definite matrix solution of the Equation (3.38) (the Lyapunov equation) with any positive definite matrix Q_i.

$$A_i^T P_i + P_i A_i = -Q_i \tag{3.38}$$

Proof. The minimum fuzzy estimate error w_i can be defined as

$$w_i = \left[\hat{\alpha}_i\left(x|\theta_i^*\right) - \alpha_i(x)\right] + \sum_{j=1}^{p=2}\left[\hat{\beta}_{ij}\left(x|\theta_{ij}^*\right) - \beta_{ij}(x)\right]u_j \tag{3.39}$$

where θ_i^* and θ_{ij}^* are the optimal values of adjustable parameters

Equation (3.34) is modified by adding and subtracting the components $\hat{\alpha}_i\left(x|\theta_i^*\right)$ and $\hat{\beta}_{ij}\left(x|\theta_{ij}^*\right)$, and then using the definition in Equation (3.39) to create the next error equation

$$\dot{e}_i = A_i e_i + b_i[w_i + \varphi_{\alpha_i}^T \xi(x) + \sum_{j=1}^{p=2}\varphi_{\beta_{ij}}^T \xi(x)u_j] \tag{3.40}$$

$\varphi_{\alpha_i} = \left(\theta_i - \theta_i^*\right)$ and $\varphi_{\beta_{ij}} = \left(\theta_{ij} - \theta_{ij}^*\right)$ are the parameter errors, and there derivatives are provided by:

$$\dot{\varphi}_{\alpha_i} = \dot{\theta}_i \tag{3.41}$$

$$\dot{\varphi}_{\beta_{ij}} = \dot{\theta}_{ij} \tag{3.42}$$

Then the following positive Lyapunov function is developed:

$$V_i = \frac{1}{2}e_i^T P_i e_i + \frac{1}{2\gamma_i}\varphi_{\alpha_i}^T \varphi_{\alpha_i} + \sum_{j=1}^{p=2}\frac{1}{2\gamma_{ij}}\varphi_{\beta_{ij}}^T \varphi_{\beta_{ij}} \tag{3.43}$$

Equation (3.43) has the following time derivative along the trajectories Equations (3.40)–(3.42):

$$\begin{aligned}\dot{V}_i = &-\frac{1}{2}e_i^T Q_i e_i + \frac{1}{\gamma_i}\varphi_{\alpha_i}^T\left(\dot{\theta}_i + \gamma_i e_i^T P_i b_i \xi(x)\right) \\ &+ \left(\frac{1}{\gamma_{ij}}\sum_{j=1}^{p=2}\varphi_{\beta_{ij}}^T \dot{\theta}_{ij} + e_i^T P_i b_i \sum_{j=1}^{p=2}\varphi_{\beta_{ij}}^T \xi(x)u_j\right) + e_i^T P_i b_i w_i\end{aligned} \tag{3.44}$$

Substituting the updating rules of the parameters in Equations (3.36) and (3.37) into Equation (3.44) yields:

Provided that

$$\dot{V}_i = -\frac{1}{2}e_i^T Q_i e_i + e_i^T P_i b_i w_i \tag{3.45}$$

$\|e_i\| \geq \frac{4\sigma_i \lambda_{max}(P_i)}{\beta_i \lambda_{min}(Q_i)} = r_i$, Equation (3.45) can be formed as

$$\dot{V}_i \leq -\frac{1}{2}(1 - \beta_i)\lambda_{min}(Q_i)\|e_i\|^2 \tag{3.46}$$

where $0 < \beta_i < 1$, $\sigma_i > 0$, such that $\|\underline{w}_i\| \leq \sigma_i$, $\lambda_{min}(Q_i)$ and $\lambda_{max}(P_i)$. We find that the tracking error is globally finally constrained by bounds $\mu_{bi} = r_i \sqrt{\frac{\lambda_{max}(P_i)}{\lambda_{min}(P_i)}}$ due to the positive definiteness of Equation (3.43) and the negative definiteness of Equation (3.46) [46].

3.2.4 PROPOSED ADAPTIVE FUZZY CONTROLLER IMPLEMENTATION FOR GCIS

Equations (3.26), (3.27), (3.32), (3.36), and (3.37), respectively, are used to apply the suggested AFC fuzzy sets F_k^i have to be chosen. To calculate the vector of FBFs provided in Equation (3.28), fuzzy sets are used. The FBFs for each state of the system are generated using three Gaussian fuzzy sets: Negative (N), Zero (Z), and Positive (P). The membership functions (MFs) characterize these fuzzy collections. The generic form of Gaussian type MFs is provided by

$$\mu_{F_k^i}(x_k) = \exp\left(-\frac{\left(x_k - \overline{x}_k^i\right)^2}{\sigma_k^i}\right) \tag{3.47}$$

where $\overline{x}_k^i$ is the center of the ith fuzzy set and σ_k^i is the width.

Figure 3.2 depicts the suggested controller's block diagram. A dq frame transformation of the grid voltage and current is obtained first, as shown in the block diagram. Unknown GCIS parameters are estimated then with the calculation starting from predetermined starting θ_i and θ_{ij} values. The control signals were then generated using AFC low in Equation (3.30). The PWM signal was obtained to operate the inverter. It's noted that the MPPT algorithm has released the signal V_{dcref}.

3.2.5 SIMULATION CASES AND RESULTS

The suggested AFC was applied and evaluated in the MATLAB®/Simulink [48] for a GCIS (see Table 3.1) to determine the effectiveness of the proposed controller. The remaining design parameters were chosen as follows: $k_{01} = 10$, $k_{02} = k_{12} = 10,000$, $\gamma_1 = 40$, $\gamma_2 = 0.01$, $\gamma_{11} = 0.01$, $\gamma_{12} = 0.1$, $\gamma_{21} = 0.1$, and $\gamma_{22} = 1$. Table 3.2 lists the parameters of the Gaussian MF for each state of the system. Figure 3.3 shows the MFs for the state x_1 as an example of state MFs.

The suggested AFC was tested with a variety of scenarios, including tracking of unity and changes of power factor (PF), and robust tracking. For all simulation instances, smooth reference values were employed.

3.2.5.1 Case I: Tracking of Unity Power Factor

Simulation was performed in this scenario by setting the reference grid current $i_{qref} = 0.0$ A. Figure 3.4 shows the output voltage and current. The grid voltage and current are in phase in this diagram, indicating unity PF functioning. The suggested controller's reactive i_q and active i_d current tracking outputs are shown in Figure 3.5(a, b). Figure 3.6 depicts the DC voltage and reference voltage.

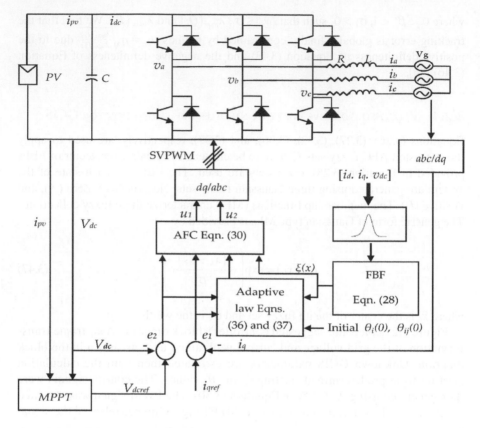

FIGURE 3.2   AFC for the grid-connected inverter system.

TABLE 3.1

Parameters of the System

Parameter	Value
Voltage of the grid rms	120 V
L	2 mH
R	0.1 Ω
Frequency of grid	50 Hz
C	2,200 μF
V_{dc}	540 V
i_{pv}	3.46 A

TABLE 3.2

Gaussian MF Parameters

State $\downarrow$	Fuzzy Set $\rightarrow$	N	Z	P
	x_1	$\overline{x}_1^N = -5$	$\overline{x}_1^Z = 0$	$\overline{x}_1^P = 5$
		$\sigma_1^N = 6$	$\sigma_1^Z = 6$	$\sigma_1^P = 6$
	x_2	$\overline{x}_2^N = -0.1$	$\overline{x}_2^Z = 0$	$\overline{x}_2^P = 0.1$
		$\sigma_2^N = 0.005$	$\sigma_2^Z = 0.005$	$\sigma_2^P = 0.005$
	x_3	$\overline{x}_3^N = 525$	$\overline{x}_3^Z = 550$	$\overline{x}_3^P = 575$
		$\sigma_3^N = 100$	$\sigma_3^Z = 100$	$\sigma_3^P = 100$

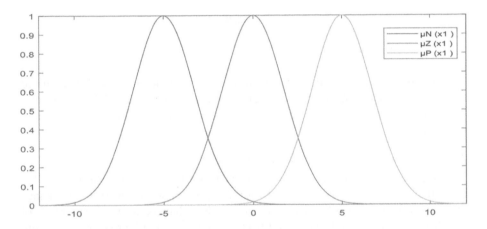

FIGURE 3.3 x_1 MFs.

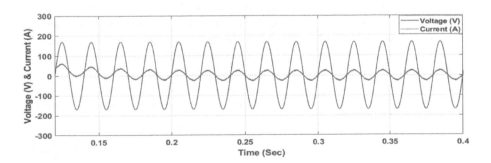

FIGURE 3.4 Voltage and current.

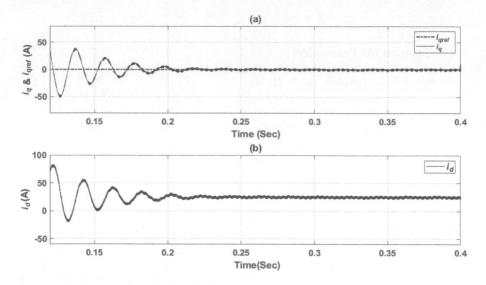

FIGURE 3.5   Current output: (a) i_q, i_{qref}; (b) i_d.

The control signals u_1 and u_2 are shown bounded in Figure 3.7(a, b). With bounded tracking error and bounded control signals, the proposed controller delivers outstanding tracking performance based on the results obtained in the case of the unity power factor.

3.2.5.2   Case II: Power Factor Tracking Changes

The suggested AFC's performance for power factor tracking was examined in this case study. The system was supposed with $i_{qref} = 0$ at first, then a 10-A step change in i_{qref} was imposed at 0.4 s. The power factor was changed to 0.937 as a result of the change in i_{qref}. The effects of modifications in i_{qref}, i_q, and i_d are shown in Figure 3.8(a, b). The findings show that i_q soon approaches its new reference. As a consequence, the collected data clearly demonstrate that the suggested AFC can track power factor changes.

Figure 3.9 depicts the output voltage and current. After $t = 0.4$ s, a phase change between current and voltage can be seen, indicating that the required power factor is being tracked. Figure 3.10(a, b) shows the active and reactive power provided to the grid

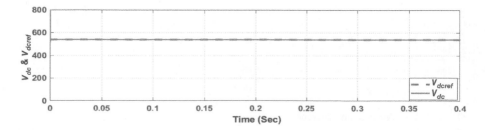

FIGURE 3.6   DC voltage and it is reference.

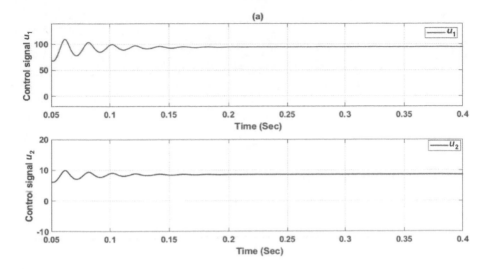

FIGURE 3.7   Signals: (a) $u_1 = v_d$; (b) $u_2 = v_q$.

by the inverter, verifying the suggested controller tracking ability. Figure 3.11(a, b) depicts the bounded control signals $u_1 = v_d$ and $u_2 = v_q$.

To assess the efficacy of AFC, the suggested controller's performance was compared with that of the PI controller as described in [49]. A step change of 10 A in i_{qref} at 0.4 s was used to perform the comparison for the power factor change tracking example. The suggested AFC and PI controller's performance is shown in Figure 3.12. Compared with the PI controller, the suggested AFC has reduced

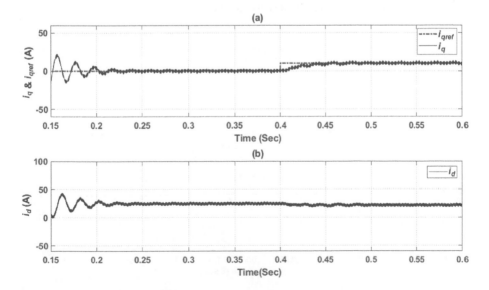

FIGURE 3.8   Current components: (a) i_q, i_{qref}; (b) i_d.

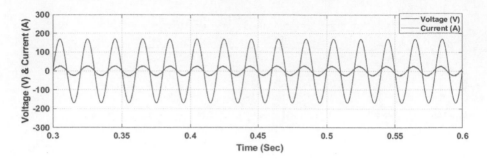

FIGURE 3.9 Voltage and current.

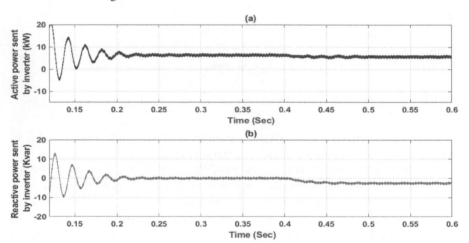

FIGURE 3.10 (a) Active power; (b) Reactive power.

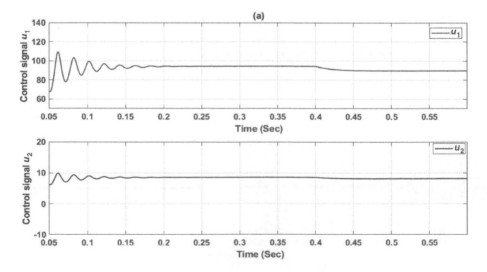

FIGURE 3.11 Control signals: (a) $u_1 = v_d$; (b) $u_2 = v_q$.

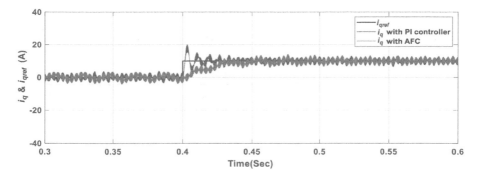

FIGURE 3.12 Comparison between PI controller and AFC.

oscillations and overshooting in terms of monitoring between i_q and i_{qref} when the step shift occurs. In addition, Table 3.3 compares the performing of the suggested controller, the PI controller, and the Takagi–Sugeno–Kang-type probabilistic fuzzy neural network (TSKPFNN) controller provided in [34]. Based on the data shown in Table 3.3, it can be concluded that the suggested AFC outperforms the PI and TSKPFNN controllers.

3.2.5.3 Case III: Robust Tracking

The GCIS parameters are sometimes time varying or not exactly defined, resulting in parametric uncertainties in which the filters connected to the grid inductance value change over time, influenced by the grid impedance value, which varies depending on grid structure and conditions, causing resonance and instability problems. Variations in ambient operating temperature, as well as changes in applied voltage and frequency, can affect DC-link capacitance values. The suggested AFC's robustness was investigated for GCIS parameter alterations in this simulated case. Various percentages of changes in the filter inductor L and the DC-link capacitor C were simulated. Figure 3.13 shows i_q with 10% variations in filter inductor L. Figure 3.14 shows the proposed controller's bounded control signals with the same L variation. The obtained findings show the AFC's robustness when the filter inductor is increased.

Simulations were conducted for 30% increase and 20% reduction in C to test the robustness of the suggested AFC for fluctuations in DC-link capacitor C.

TABLE 3.3

Comparison of the Proposed Controller's Performance with That of the PI and TSKPFNN Controllers

Controller	Max Overshoot (%)	Settling Time (S)
PI	75	0.04
TSKPFNN	12.24	0.3
Suggested AFC	0.0	0.035

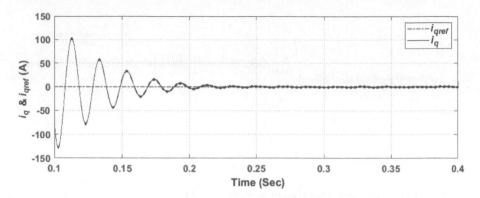

FIGURE 3.13   Ten percent increase in L: i_q and i_{qref}.

Figures 3.15–3.18 depict the performance of the GCIS with the suggested AFC and implemented changes, with Figure 3.15(a, b) displaying i_q and i_d with a 30% increase in C. Figure 3.16 shows grid voltage and current with the same rise in C. Figure 3.17 depicts the constrained control signals in the event of a 20% drop in C. Figure 3.18 depicts the tracking between i_q and i_{qref}. The controller's robustness is demonstrated by the simulation results produced with modifications in C.

Figure 3.19 also shows how the suggested controller performs when the inductor and capacitor are both changed at the same time with a 10% rise in L. The performance of the GCIS demonstrates that the suggested AFC can deal with the uncertainty of the GCIS parameters and achieve the necessary tracking performance based on all simulation results for parameter uncertainties.

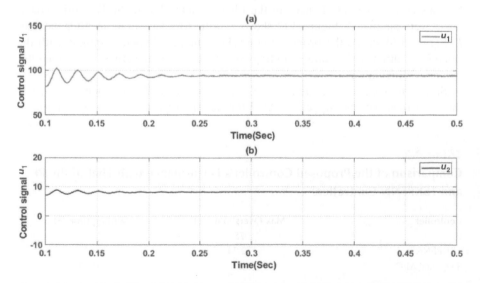

FIGURE 3.14   Ten percent increase in L. Control signals: (a) $u_1 = v_d$; (b) $u_2 = v_q$.

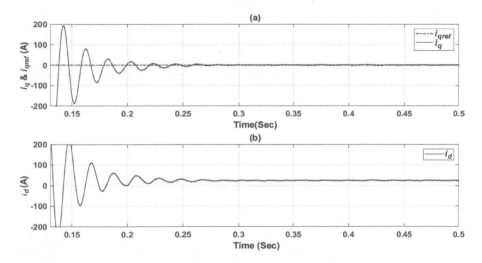

FIGURE 3.15 Thirty percent increase in C: (a) i_q, i_{qref}; (b) i_d.

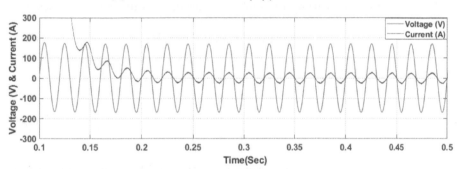

FIGURE 3.16 Thirty percent increase in C.

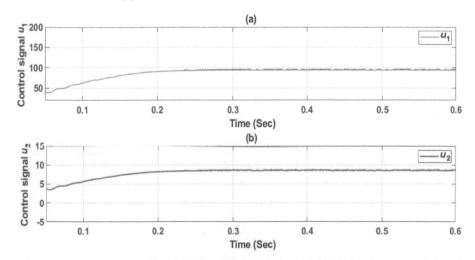

FIGURE 3.17 Twenty percent decrease in C: (a) $u_1 = v_d$; (b) $u_2 = v_q$.

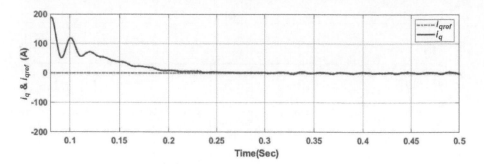

FIGURE 3.18   Twenty percent decrease in C: i_q and i_{qref}.

3.2.5.4   Case IV: Model Uncertainty

In the existence of modeling errors that are inherent in the nature of the GCIS, the suggested AFC [Equation (3.30)] can accomplish tracking. The inclusion of the PV in the GCIS model given by Equation (3.10) is the major source of modeling errors, and Equation (3.10) may be recast as follows to account for these uncertainties:

$$\dot{x} = \left(f(x) + \Delta f(x)\right) + g(x)u \tag{3.48}$$

where $\Delta f(x)$ given by

$$\Delta f = \begin{bmatrix} \Delta f_1 \\ \Delta f_2 \\ \Delta f_3 \end{bmatrix} \tag{3.49}$$

The function $\alpha(x)$ in Equation (3.17) that comes from feedback linearization will be affected in this case by $\Delta\alpha(x)$ provided by

$$\Delta\alpha(x) = \begin{bmatrix} \Delta f_2 \\ \Delta\alpha_2 \end{bmatrix} \tag{3.50}$$

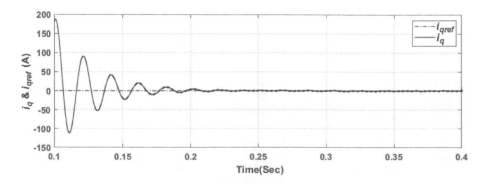

FIGURE 3.19   Ten percent increase in L and 30% increase in C: i_q and i_{qref}.

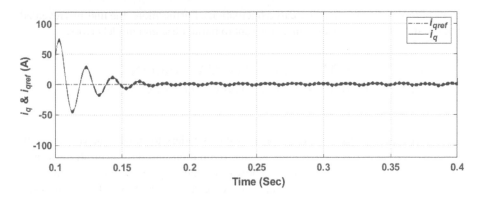

FIGURE 3.20 Modeling uncertainty $\Delta f_3 = 5\%$: i_q and i_{qref}.

In Equation (3.50), $\Delta\alpha_2$ is given by

$$\Delta\alpha_2 = -\frac{1}{Cx_3}\left(v_{gd}\Delta f_1 + v_{gq}\Delta f_2\right) + n\ \Delta f_3 \qquad (3.51)$$

where $n = \frac{\left(v_{gd}x_1 + v_{gq}x_2\right)}{Cx_3^2}$.

We assumed that there is an uncertainty $\Delta f_3 = 5\%$, which is mostly due to the existence of the PV current, to evaluate the performance of the suggested AFC with modeling uncertainty. Other $\Delta f_1 = \Delta f_2$ uncertainties were believed to be zero. Figure 3.20 depicts the simulation result for tracking i_{qref}. It can be observed that the suggested AFC controller can follow the reference reactive current i_{qref} and keep the GCIS functioning at unity power factor even when modeling uncertainty exists.

3.2.6 CONCLUSIONS

This chapter investigates grid-connected inverter control. An AFC technique for GCIS is presented to overcome the nonlinearity and uncertainty concerns of GCIS. The designed controller is based on the input-output feedback linearization concept and the fuzzy system's estimate capabilities. A nonlinear MIMO system is used to model the GCIS. The AFC law is developed by approximating the unknown nonlinear functions that arise in the input-output linearizing model using a fuzzy system.

The suggested controller is robust against parametric uncertainties due to its ability to estimate unknown parameters for various operating circumstances. The output tracking error is shown to be globally ultimately bounded by establishing closed-loop stability utilizing the Lyapunov function analysis. The suggested AFC was built for GCIS and evaluated in the MATLAB/Simulink for multiple operating situations such as unity power factor tracking, tracking of power factor variations, and robust tracking to test the efficacy of the proposed technique. With bounded tracking error and bounded control inputs, the proposed AFC delivers outstanding tracking performance, according to simulation findings. In the instance of power factor change tracking, the suggested AFC outperformed the PI control and TSKPFNN controller in terms of responsiveness and decreased fluctuations. Furthermore, the findings revealed that the suggested AFC is extremely tolerant to parametric and model errors.

3.3 ADAPTIVE INTERVAL TYPE 2 FUZZY TRACKING CONTROL OF PV GRID-CONNECTED INVERTERS

3.3.1 INTRODUCTION

It's not straightforward to develop successful controllers for nonlinear systems with unknown characteristics. An adaptive approximation-based control strategy is a practical answer for dealing with unknown nonlinear systems. Based on the estimate property of FLSs, AFC techniques have been suggested [37, 50–52] and used for several power systems and drive applications [38–41, 53] to control uncertain nonlinear systems. Type 1 FLSs (T1FLSs) and their applications were first the focus of research. The T1FLSs provide significant control capabilities in a variety of applications, including incorporating uncertainty in cases when MFs for a fuzzy group may be determined accurately by a single numerical value.

However, there are more difficult circumstances where determining the precise numerical value of the MFs for unknown systems is exceedingly difficult. Because the information required is ambiguous, creating the fuzzy rule base that is employed in fuzzy systems is a tough operation. As a result, the use of inappropriate fuzzy rules in MFs may cause misunderstanding. Then, despite the T1FLSs' usefulness, other circumstances may arise in which they are unable to achieve the required degree of precision or efficiency. To address this flaw, Zadeh [54] in 1975 presented the type 2 fuzzy logic system (T2FLS), followed by Mendel et al. [55–57] with T2FLCs.

Because the degrees of MFs are crisp, the T1FLS' ability to cope with a large amount of uncertainty is restricted. As illustrated in Figure 3.21 [58], the MFs employed in T2FLS are a mixture of T1FLS two MFs: the upper MF (UMF) and the lower MF (LMF). Then as a result, the T2FLS MF is transformed into a three-dimensional (3D) model with an additional degree of freedom. The "footprint of uncertainty" (FOU) is the function value at each location in a two-dimensional space, and it is the third dimension. Figure 3.21 shows that the MF can have many

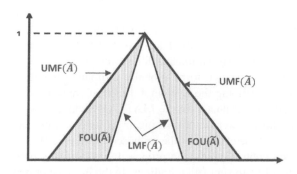

FIGURE 3.21   T2FLS MFs.

responses to the same input variable [59]. T2FLS operates the same as T1FLS without the uncertainty [58].

In a variety of applications T2FLS has been used and has shown to be effective in dealing with increased degrees of uncertainty [60]. Image processing [61] and vehicle categorization [62] are examples of these uses. T2FLS has also been shown to be effective in medicine [25].

T2FLC has been used in a variety of applications [29–32, 63–67]. To regulate a DC-DC converter the T2FLC is used, and the findings show that type 2 performs better than type 1. In [30], based MPPT the IT2FLC approach is presented for solar systems. The work in [31] recommended utilizing a T2FLC to implement the MPPT to address the rules' uncertainties when there are severe weather conditions and changes.

For a single-phase grid-connected PV system, the IT2FLC is employed as the MPPT algorithm [32]. The T2FLC [63] was proposed in the realm of hybrid electric cars to handle energy management efficiently. To tackle unanticipated internal disturbances and data uncertainties, the direct and indirect adaptive T2FLC was presented in [64, 65]. In [68, 69], a proportional-derivative (PD) type 2 fuzzy neural network (T2FNN) and T2FNN are offered to control the DC-link voltage of a three-phase inverter. References [70, 71] offer an adaptive fuzzy controller for GCIS based on type 1 fuzzy. The authors expect that using IT2FLS will considerably improve the tracking performance of the controller described in [70]. Furthermore, there is no published work that covers the type 2 fuzzy adaptive fuzzy controller for theGCIS topic, to the best of the authors' knowledge. For a PV GCIS, an adaptive interval type 2 fuzzy-based tracking control (AIT2FAC) is developed in this study. GCISs are very nonlinear and unreliable. If nonlinearities and uncertainties aren't handled by a fast-acting inverter controller, they might result in system issues. To estimate nonlinearities and adjust for the uncertainties inherited in PV systems, the type 2 fuzzy system is used. The type 2 fuzzy controller is suggested in this study to better accurately handle the system uncertainty given in [70]. A comparative study was done between both controllers to determine the superiority of the suggested controller over the T1FLS in [70].

The contributions of this chapter section are summarized as follows:

- For GCIS, the study presents an interval type 2 fuzzy-based tracking control.
- The implementation of the suggested controller does not need prior knowledge of the system's mathematical model.
- Under a variety of operating situations, the suggested controller achieves outstanding tracking efficiency.
- The suggested controller surpasses the T1FLS controller [70].

The remainder of the chapter section is laid out as follows. The IT2FLS is shown in Section 3.3.2. The suggested AIT2FAC for nonlinear systems is discussed in Section 3.3.3. The AIT2FAC design for a PV GCIS is shown in Section 3.3.4. In Section 3.3.5, the simulation results are reported and compared with the T1FLS adaptive controller. Section 3.3.6 is where the conclusions are reached.

3.3.2 INTERVAL TYPE 2 FUZZY LOGIC SYSTEMS

Because T2FLSs are mathematically complicated, a simplified variant known as interval type 2 fuzzy logic systems (IT2FLS) has been developed [72]. In the universal set X, a type 2 fuzzy set is specified as $\tilde{A}$, which is defined by a type 2 MF $\mu_{\tilde{A}}(x)$ as in Equation (3.52) [64, 72]:

$$\tilde{A} = \int_{x\in X} \mu_{\tilde{A}}(x)x = \int_{x\in X}\left[\int_{u\in J_x\subseteq[0,1]} f_x(u)/u\right]/x \tag{3.52}$$

where $\mu_{\tilde{A}}(x)$ is defined as a secondary MF in $[0,1]$, and $f_x(u)$ is defined as a secondary MF amplitude, where $0\le f_x(u)\le 1$. The primary membership of x is the secondary MF domain. In (1), J_x is the primary membership of x and u is a fuzzy set in [0,1], and not a crisp point in [0,1]. In $f_x(u)=1, \forall\ u\in J_x\subseteq[0,1]$, $\mu_{\tilde{A}}(x)$, it's also known as an interval type 2 MF. Equation (3.52), on the other hand, may be written as [64].

$$\tilde{A} = \int_{x\in X} \mu_{\tilde{A}}(x)x = \int_{x\in X}\left[\int_{u\in J_x\subseteq[0,1]} 1/u\right]/x \tag{3.53}$$

In the FOU area, which is surrounded by a UMF defined as $\bar{\mu}_{\tilde{A}}(x)$ and an LMF the type 2 fuzzy set is indicated as $\underline{\mu}_{\tilde{A}}(x)$ [56]. Equation (3.53) can be formed as

$$\tilde{A} = \int_{x\in X}\left[\int_{u\in\left[\underline{\mu}_{\tilde{A}}(x),\bar{\mu}_{\tilde{A}}(x)\right]} 1/u\right]/x \tag{3.54}$$

As illustrated in Figure 3.22 [72], T2FLS has a rule base, a fuzzifier, an inference engine, and an output processor, just as T1FLS. The inference engine uses a set of rules to create a mapping from input T2FLSs to output T2FLSs, and a type reduction provides a type 1 fuzzy output [72].

Centroid, height, modified height, and center of sets, are some of the type-reduction strategies that have been presented [57]. The center of the sets type reduction will be used in this work.

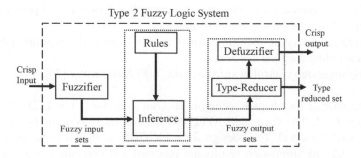

FIGURE 3.22 Type 2 FLS.

The type 2 fuzzy rule base for T2FLS with p inputs and one output $y \in Y$ may be characterized as [64]

$$Y_{cos}\left(Y^1,\ldots,Y^M,F^1,\ldots,F^M\right)=\left[y_l,y_r\right]=\int_{y^1}\cdots\int_{y^M}\int_{f^1}\cdots\int_{f^M}1/\frac{\sum_{i=1}^M f^i y^i}{\sum_{i=1}^M f^i} \qquad (3.55)$$

where M is the number of If–Then rules, Y_{cos} is the interval set specified by two end points y_l and y_r, $y^i \in Y^i = \left[y_l^i, y_r^i\right]$, F^i is the antecedent type 2 sets, and $f^i \in F^i = \left[\underline{f}^i, \overline{f}^i\right]$. IT2FLS with singleton fuzzification, $\underline{f}^i$ and $\overline{f}^i$, can be found as

$$\underline{f}^i = \underline{\mu}_{\tilde{F}_1^i}(x_1)*\cdots*\underline{\mu}_{\tilde{F}_p^i}(x_p) \qquad (3.56)$$

$$\overline{f}^i = \overline{\mu}_{\tilde{F}_1^i}(x_1)*\cdots*\overline{\mu}_{\tilde{F}_p^i}(x_p) \qquad (3.57)$$

A consequence type 2 interval set centroid is $y^i \in Y^i$, and for any value $y \in Y_{cos}$, y can be stated as

$$y = \frac{\sum_{i=1}^M f^i y^i}{\sum_{i=1}^M f^i} \qquad (3.58)$$

y_l and y_r can be stated as an FBF ξ as [57, 64]

$$y_l = \frac{\sum_{i=1}^M f_l^i y_l^i}{\sum_{i=1}^M f_l^i} = \sum_{i=1}^M y_l^i \xi_l^i = \underline{y_l}^T \underline{\xi_l} \qquad (3.59)$$

$$y_r = \frac{\sum_{i=1}^M f_r^i y_r^i}{\sum_{i=1}^M f_r^i} = \sum_{i=1}^M y_r^i \xi_r^i = \underline{y_r}^T \underline{\xi_r} \qquad (3.60)$$

where $\xi_l^i = \frac{f_l^i}{\sum_{i=1}^M f_l^i}$, $\xi_r^i = \frac{f_r^i}{\sum_{i=1}^M f_r^i}$, $\underline{\xi_l} = \left[\xi_l^1,\ldots,\xi_l^M\right]$, $\underline{\xi_r} = \left[\xi_r^1,\ldots,\xi_r^M\right]$, $\underline{y_l}^T = \left[y_l^1,\ldots,y_l^M\right]$ and $\underline{y_r}^T = \left[y_r^1,\ldots,y_r^M\right]$.

Lastly, the defuzzified crisp value from is found as [64]

$$y(x) = \frac{y_l + y_r}{2} = \underline{\xi}^T \Theta \qquad (3.61)$$

where $\underline{\xi}^T = \frac{1}{2}\left[\underline{\xi_r}^T \ \underline{\xi_l}^T\right]$, $\Theta^T = \left[\Theta_r^T \ \Theta_l^T\right]$, $\Theta_r^T = \left[\underline{y_r} \ \overline{y_r}\right]$ and $\Theta_l^T = \left[\underline{y_l} \ \overline{y_l}\right]$.

3.3.3 Design of Adaptive Interval Type 2 Fuzzy Approximation Controller

Take into consideration the MIMO system

$$\dot{x} = f(x) + g(x)u \tag{3.62}$$

The system's unknown nonlinear functions are $f(x)$ and $g(x)$. The following are the primary assumptions that the study is based on.

> *Assumption 1:* System Equation (3.62) is a MIMO system with an equal number of inputs and outputs.
> *Assumption 2:* The state vectors may be calculated.
> *Assumption 3:* The reference tracking signals $y_{ref1}, \cdots y_{refp}$, as well as their derivatives, are smooth and constrained.
> *Assumption 4*: The no singularity of the system matrix $\beta(x)$.

The input-output feedback linearizing control for a typical MIMO nonlinear system may be stated

$$u = \beta^{-1}(x)\left[y^{(r)} - \alpha(x)\right] \tag{3.63}$$

where $y^{(r)} = \left[y_1^{(r_1)} \cdots y_p^{(r_p)}\right]^T$; $u = \left[u_1 \cdots u_p\right]^T$; $r_1, r_2 \cdots r_p$ are the relative degrees of the outputs $y_1, y_2 \cdots y_p$, respectively; p is the number of both inputs and outputs; $\alpha(x) \in R^{p \times 1}$; and $\beta(x) \in R^{p \times p}$ [70].

To use control in Equation (3.63), you'll need to know the exact values of $\alpha_i(x)$ and $\beta_{ij}(x)$. In practice, however, those parameters may be unknown or wrong. To estimate $\alpha_i(x)$ and $\beta_{ij}(x)$, the IT2FLS is used where the predicted functions $\hat{\alpha}_i(x)$ and $\hat{\beta}_{ij}(x)$ are constructed using the FBF expansion $\xi(x)$ obtained using IT2FLS as

$$\hat{\alpha}_i(x) = \theta_i^T \xi(x) \tag{3.64}$$

$$\hat{\beta}_{ij}(x) = \theta_{ij}^T \xi(x) \tag{3.65}$$

where θ_i, θ_{ij} depict vectors of variables that are changeable.

We get the input-output linearizing control for a general MIMO nonlinear system by substituting $\alpha(x)$ and $\beta(x)$ with their corresponding fuzzy estimates in the input-output linearizing control.

$$y^{(r)} = \hat{\alpha}(x) + \hat{\beta}(x)u \tag{3.66}$$

where $\hat{\alpha}(x) = \begin{bmatrix} \hat{\alpha}_1 \\ \vdots \\ \hat{\alpha}_p \end{bmatrix}$ and $\hat{\beta}(x) = \begin{bmatrix} \hat{\beta}_{11} & \cdots & \hat{\beta}_{1p} \\ \vdots & \vdots & \vdots \\ \hat{\beta}_{p1} & \cdots & \hat{\beta}_{pp} \end{bmatrix}$.

After that, the linearizing control rule may be constructed using fuzzy estimates $\hat{\alpha}(x)$ and $\hat{\beta}(x)$ as

$$u = \hat{\beta}^{-1}(x)\left(v - \hat{\alpha}(x)\right) \tag{3.67}$$

and,

$$v = \begin{bmatrix} v_1 \\ \vdots \\ v_p \end{bmatrix} = y_{ref}^r + K\underline{e} \tag{3.68}$$

where $\quad y_{ref}^r = \left[y_{ref1}^{(r_1)} \cdots y_{refp}^{(r_p)} \right]^T, \quad K = diag\left[\underline{k_1} \;\; \underline{k_2} \;\cdots\; \underline{k_p} \right], \quad \underline{e} = \left[\underline{e_1} \;\; \underline{e_2} \;\cdots\; \underline{e_p} \right]^T,$
$\underline{k_j} = \left[k_{0i} \;\; k_{1i} \cdots k_{(n-1)I} \right]$, and $\underline{e_j} = \left[e_I \;\; \dot{e}_I \cdots e_i^{(n-1)} \right]$ and $e_i = y_{refi} - y_I$ [70].

3.3.3.1 Closed-Loop Stability

The error equation may be constructed using Equations (3.67) and (3.68).

$$\begin{bmatrix} e_1^{r_1} \\ \vdots \\ e_p^{r_p} \end{bmatrix} = -K\underline{e} + \left(\hat{\alpha}(x) - \alpha(x)\right) + \left(\hat{\beta}(x) - \beta(x)\right)u \tag{3.69}$$

The ith error equation may be stated as

$$e_i^{r_i} = -\underline{k}_i \underline{e}_i + \Delta\alpha_i(x) + \sum_{j=1}^{p} \Delta\beta_{ij}(x)u_j \tag{3.70}$$

and Equation (3.70) turns into

$$\dot{\underline{e}} = A_i \underline{e}_i + \left[\Delta\alpha_i(x) + \sum_{j=1}^{p} \Delta\beta_{ij}(x)u_j \right] b_i \tag{3.71}$$

where $A_i = \begin{bmatrix} 0 & 1 & 0 & \cdots & 0 \\ 0 & 0 & 1 & \cdots & 0 \\ \vdots & \vdots & \vdots & \cdots & \vdots \\ 0 & 0 & 0 & 0 & 1 \\ -k_{(n-1)I} & k_{(n-2)i} & \cdots & \cdots & -k_{0i} \end{bmatrix}$ and $b_i = \begin{bmatrix} 0 \\ 0 \\ \cdots \\ 0 \\ 1 \end{bmatrix}$.

The minimal fuzzy estimate error w_i in terms of the ideal values of the adjustable parameters θ_i^* and θ_{ij}^* [37] is

$$w_i = \left[\hat{\alpha}_i\left(x|\theta_i^*\right) - \alpha_i(x) \right] + \sum_{j=1}^{p} \left[\hat{\beta}_{ij}\left(x|\theta_{ij}^*\right) - \beta_{ij}(x) \right] u_j \tag{3.72}$$

So Equation (3.71) becomes

$$\dot{e} = A_i \underline{e}_i + b_i [w_i + \varphi_{\alpha_i}^T \xi(x) + \sum_{j=1}^{p} \varphi_{\beta_{ij}}^T \xi(x) u_j] \tag{3.73}$$

where $\varphi_{\alpha_i} = \left(\theta_i - \theta_i^*\right)$, $\varphi_{\beta_{ij}} = \left(\theta_{ij} - \theta_{ij}^*\right)$ and $\dot{\varphi}_{\alpha_i} = \dot{\theta}_i$, $\dot{\varphi}_{\beta_{ij}} = \dot{\theta}_{ij}$ are the parameter errors and its derivatives, respectively [70].

The next positive Lyapunov function is created to analyze closed-loop stability and identify the updating laws of the adjustable parameters θ_i and θ_{ij}.

$$V_i = \frac{1}{2} \underline{e}_i^T P_i \underline{e}_i + \frac{1}{2\gamma_i} \varphi_{\alpha_i}^T \varphi_{\alpha_i} + \sum_{j=1}^{p} \frac{1}{2\gamma_{ij}} \varphi_{\beta_{ij}}^T \varphi_{\beta_{ij}} \tag{3.74}$$

where

$$A_i^T P_i + P_i A_i = -Q_I \tag{3.75}$$

Equation (3.74) has a time derivative of

$$\dot{V}_i = -\frac{1}{2} \underline{e}_i^T Q_i \underline{e}_i + \frac{1}{\gamma_i} \varphi_{\alpha_i}^T \left(\dot{\theta}_i + \gamma_i \underline{e}_i^T P_i b_i \xi(x) \right)$$
$$+ \left(\frac{1}{\gamma_{ij}} \sum_{j=1}^{p} \varphi_{\beta_{ij}}^T \dot{\theta}_{ij} + \underline{e}_i^T P_i b_i \sum_{j=1}^{p} \varphi_{\beta_{ij}}^T \xi(x) u_j \right) + \underline{e}_i^T P_i b_i w_i \tag{3.76}$$

by deciding on the update laws for the parameters as

$$\dot{\theta}_i = -\gamma_i \underline{e}_i^T P_i b_i \xi(x) \tag{3.77}$$

$$\dot{\theta}_{ij} = -\gamma_{ij} \underline{e}_i^T P_i b_i \xi(x) u_j \tag{3.78}$$

Then, by replacing the updating rules of the parameters, Equation (3.76) becomes

$$\dot{V}_i = -\frac{1}{2} \underline{e}_i^T Q_i \underline{e}_i + \underline{e}_i^T P_i b_i w_i \tag{3.79}$$

It is possible to obtain closed-loop stability by employing Equation (3.79). This may be demonstrated as follows. Assume that the smallest fuzzy approximation error has a norm of $\|w_i\| \leq \sigma_i$ and taking the norm of Equation (3.79) to have

$$\dot{V}_i \leq -\frac{1}{2} \lambda_{min}(Q_i) \|e_i\|^2 + \sigma_i \lambda_{max}(P_i) \|e_i\| \tag{3.79a}$$

where $\lambda_{min}(Q_i)$ and $\lambda_{max}(P_i)$ are the corresponding matrices' minimum and maximum cigenvalues and $\|.\|$ stands for the Euclidean norm [71]. Revising Equation(3.79a) such as

$$\dot{V}_i \leq -\frac{1}{2}(1 - \beta_i) \lambda_{min}(Q_i) \|e_i\|^2 + \left[\frac{\beta_i}{2} \lambda_{min}(Q_i) \|e_i\| - 2\sigma_i \lambda_{max}(P_i) \right] \|e_i\| \tag{3.79b}$$

where $0 < \beta_i < 1$. Given that

$$\|\underline{e}\|_i \geq \frac{4\sigma_i \lambda_{max}(P_i)}{\beta_i \lambda_{min}(Q_i)} = r_i \qquad (3.79c)$$

then Equation (3.79b) turns into

$$\dot{V_i} \leq -\frac{1}{2}(1-\beta_i)\lambda_{min}(Q_i)\|\underline{e}_i\|^2 \qquad (3.79d)$$

We infer that the tracking error is globally finally constrained due to the positive definiteness of Equation (3.74) and the negative definiteness of Equation (3.79d), and the ultimate bound μ_{bi} may be obtained as [16, 73]

$$\mu_{bi} = r_i \sqrt{\frac{\lambda_{max}(P_i)}{\lambda_{min}(P_i)}} \qquad (3.79e)$$

3.3.4 IMPLEMENTATION OF ADAPTIVE INTERVAL TYPE 2 FUZZY APPROXIMATION CONTROLLER FOR GCIS

The system's state model may be expressed as

$$\dot{x} = \begin{bmatrix} \omega x_2 - \dfrac{1}{L}x_3 \\[2mm] -\omega x_1 - \dfrac{1}{L}x_4 \\[2mm] \omega x_4 + \dfrac{1}{C}x_1 \\[2mm] -\omega x_3 + \dfrac{1}{C}x_2 \\[2mm] -\dfrac{3}{2C_{dc}x_5}(v_{gd}i_{gd} + v_{gq}i_{gq}) \end{bmatrix} + \begin{bmatrix} \dfrac{1}{L} \\[2mm] 0 \\ 0 \\ 0 \\ 0 \end{bmatrix} u_1 + \begin{bmatrix} 0 \\[2mm] \dfrac{1}{L} \\ 0 \\ 0 \\ 0 \end{bmatrix} u_2 + \begin{bmatrix} 0 \\ 0 \\ -\dfrac{1}{C}i_{gd} \\[2mm] -\dfrac{1}{C}i_{gq} \\[2mm] \dfrac{i_{pv}}{C_{dc}} \end{bmatrix} \qquad (3.80)$$

where $x \in R^{n=5}$, and control inputs u_1 and u_2 are defined by

$$x = \begin{bmatrix} x_1 \\ x_2 \\ x_3 \\ x_4 \\ x_5 \end{bmatrix} = \begin{bmatrix} i_d \\ i_q \\ v_{cd} \\ v_{cq} \\ v_{dc} \end{bmatrix} \qquad (3.81)$$

$$u = \begin{bmatrix} u_1 \\ u_2 \end{bmatrix} = \begin{bmatrix} v_d \\ v_q \end{bmatrix} \qquad (3.82)$$

The purpose of this research is to construct an AIT2FAC based on feedback linearization, with $y = \begin{bmatrix} i_d & i_q \end{bmatrix}^T = \begin{bmatrix} y_1 & y_2 \end{bmatrix}^T$ defining the inverter dq current components will track particular reference current components $y_{ref} = \begin{bmatrix} i_{dref} & i_{qref} \end{bmatrix}^T = \begin{bmatrix} y_{ref1} & y_{ref2} \end{bmatrix}^T$.

The MIMO model of the GCIS in Equation (3.80) may be transformed to a feedback linearizable form [46]. The grid-connected system's relative degree is denoted by $r_1 = r_2 = 1$, and the feedback linearizing control rule is denoted by

$$\begin{bmatrix} u_1 \\ u_2 \end{bmatrix} = \beta^{-1}(x) \begin{bmatrix} v_1 - \alpha_1 \\ v_2 - \alpha_2 \end{bmatrix} \tag{3.83}$$

where

$$\alpha(x) = \begin{bmatrix} \alpha_1 \\ \alpha_2 \end{bmatrix} = \begin{bmatrix} L_f h_1(x) \\ L_f h_2(x) \end{bmatrix} = \begin{bmatrix} \omega x_2 - \dfrac{1}{L} x_3 \\ -\omega x_1 - \dfrac{1}{L} x_4 \end{bmatrix} \tag{3.84}$$

$$\beta(x) = \begin{bmatrix} L_{g_1} h_1 & L_{g_2} h_1 \\ L_{g_1} h_2 & L_{g_2} h_2 \end{bmatrix} = \begin{bmatrix} \dfrac{1}{L} & 0 \\ 0 & \dfrac{1}{L} \end{bmatrix} \tag{3.85}$$

and v_1 and v_2 are the new input vector.

Given that $r_1 + r_2 = 2 = m < n$, the inverter current control mechanism for grid-connected inverters is partially linearizable. It is possible to verify the stability of system zero dynamics [74].

In the PV GCIS depicted in Figure 3.22, the suggested AIT2FAC is used. Figure 3.23 depicts the suggested controller's block diagram. As indicated in the block diagram, the FBF is computed using system states and interval type-2 Gaussian MF. The control rules [Equations (3.77) and (3.78)] were used to estimate the unknown parameters of the PV GCIS, with the computation beginning from specified initial values of $\theta_i(0)$ and $\theta_{ij}(0)$. Then, the control signals were acquired using AFC law [Equation (3.67)], and PWM signals were created. The AIT2FAC law [Equation (3.67)] for the controlled system may be stated as

$$\begin{bmatrix} u_1 \\ u_2 \end{bmatrix} = \begin{bmatrix} \hat{\beta}_{11}(x) & \hat{\beta}_{12}(x) \\ \hat{\beta}_{21}(x) & \hat{\beta}_{22}(x) \end{bmatrix}^{-1} \begin{bmatrix} -\hat{\alpha}_1(x) + v_1 \\ -\hat{\alpha}_2(x) + v_2 \end{bmatrix} \tag{3.86}$$

The fuzzy approximation seen in Equations (3.64) and (3.65) as well as the update rules seen in Equations (3.77) and (3.78) are used to generate $x\,\hat{\alpha}_1$ and $\hat{\beta}_{ij}$. A_i, b_i, $i = 1,2$ in Equation (3.71) are provided by $A_1 = -k_{01}$, $A_2 = -k_{02}$ and $b_1 = b_2 = 1$, where k_{01} and k_{02} are chosen so A_1 and A_2 are precisely Hurwitz matrices. Furthermore, because adaptation methods are iterative, the selected gains should be highly sufficient to provide stability at startup. Practical limitations and a trial-and-error method are used to determine the design parameters γ_i and γ_{ij}.

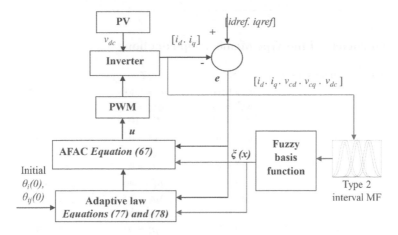

FIGURE 3.23   Block diagram of the proposed controller.

An interval type 2 Gaussian MF in Equation (3.87) with an uncertain mean is in $[\mu - \Delta\mu, \ \mu + \Delta\mu]$, and a fixed standard deviation σ is used to implement the suggested AIT2FAC.

The corresponding MF's center is selected as $x_i(0)$, and the rest of the parameters in the constraint sets are selected at arbitrary in the constraint sets [37]. Three Gaussian fuzzy sets are used to produce the FBF, namely, negative (N), zero (Z), and positive (P) for each of the system states. Figure 3.24 depicts the MF for the state x_1, as an example of state's interval type 2 MF.

$$u_{\tilde{A}}(x) = \exp\left[-\frac{1}{2}\left(\frac{x-m}{\sigma} \right)^2 \right], \ m \in [\mu - \Delta\mu, \mu + \Delta\mu] \tag{3.87}$$

Table 3.4 lists the parameters of the MFs.

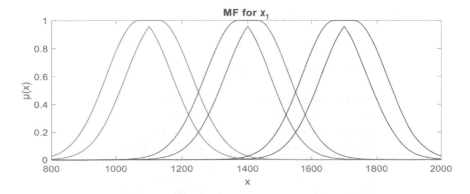

FIGURE 3.24   MFs for x_1.

TABLE 3.4
Parameters of the Type Membership Functions

State ↓	Fuzzy set →	N	Z	P
x_1		$\mu_1^N = 1,100$	$\mu_1^Z = 1,400$	$\mu_1^N = 1,700$
		$\Delta\mu_1^N = 30$	$\Delta\mu_1^Z = 30$	$\Delta\mu_1^N = 30$
		$\sigma_1^N = 100$	$\sigma_1^Z = 100$	$\sigma_1^N = 100$
x_2		$\mu_2^N = -10$	$\mu_2^Z = 0$	$\mu_2^N = 10$
		$\Delta\mu_2^N = 1$	$\Delta\mu_2^Z = 1$	$\Delta\mu_2^N = 1$
		$\sigma_2^N = 4$	$\sigma_2^Z = 4$	$\sigma_2^N = 4$
x_3		$\mu_3^N = 320$	$\mu_3^Z = 340$	$\mu_3^N = 360$
		$\Delta\mu_3^N = 2.5$	$\Delta\mu_3^Z = 2.5$	$\Delta\mu_3^N = 2.5$
		$\sigma_3^N = 8$	$\sigma_3^Z = 8$	$\sigma_3^N = 8$
x_4		$\mu_4^N = -10$	$\mu_4^Z = 0$	$\mu_4^N = 10$
		$\Delta\mu_4^N = 1$	$\Delta\mu_4^Z = 1$	$\Delta\mu_4^N = 1$
		$\sigma_4^N = 4$	$\sigma_4^Z = 4$	$\sigma_4^N = 4$
x_5		$\mu_5^N = 1,200$	$\mu_5^Z = 1,290$	$\mu_5^N = 1,380$
		$\Delta\mu_5^N = 10$	$\Delta\mu_5^Z = 10$	$\Delta\mu_5^N = 10$
		$\sigma_5^N = \sqrt{1,000}$	$\sigma_5^Z = \sqrt{1,000}$	$\sigma_5^N = \sqrt{1,000}$

3.3.5   SIMULATION RESULTS

A three-phase GCIS with the specifications provided in Table 3.5 [43] is simulated using MATLAB/SIMULINK for various operating situations to test the performance of the proposed controller.

Simulations were performed for $i_{qref} = 0$ A and $i_{dref} = 1,400$ A to confirm the unity power factor tracking. Figure 3.25 depicts the reaction of GCIS with the proposed AIT2AFC. The tracking i_d to its reference i_{dref} and the tracking of i_q to its reference i_{qref} are depicted in Figures 3.25 (a) and (b), respectively.

In the case of unity PF control, the results show that the suggested AIT2FAC offers good active and reactive current tracking performance. Furthermore, Figure 3.25(c) depicts grid current and voltage, and the voltage and current are in phase, suggesting that the PF performance is unity.

TABLE 3.5
Parameters of the System

Parameter of the System	Value
Voltage of the grid rms	415 V
Frequency of the grid	50 Hz
Components of the filter	$L = 100\ \mu H$, $C = 369\ \mu F$
DC-link capacitor	$5,000\ \mu F$
V_{dc}	1,290 V
Coefficients k_{ij}	$k_{01} = 1,000$, $k_{02} = 1,000$

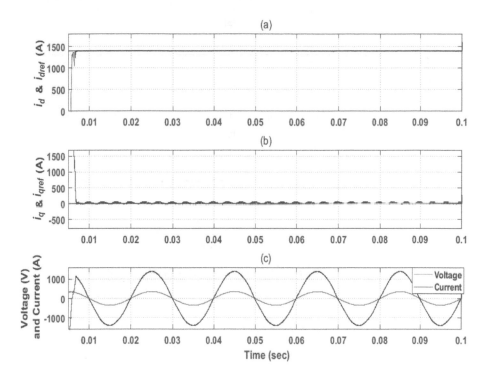

FIGURE 3.25 (a) i_d and i_{dref}; (b) i_q and i_{qref}; (c) voltage and current waveforms with power factor control to unity.

A change in i_{qref} of 400 A at t = 0.1 s is expected to show the system's performance under power factor change. Figure 3.26 depicts the reaction of the system. The grid current components i_q and i_{qref} are shown in Figure 3.26(b), and the findings reveal that i_q quickly approaches its new reference. Figure 3.27 illustrates a 1-msec phase shift between voltage and current, indicating that the required 0.988 power factor is being maintained. Figure 3.28 also depicts the change in power factor as i_{qref} changes. As a result of the simulation results, the suggested AIT2FAC can track the target power factor via changes in i_{qref}.

The suggested controller was examined with a variation in i_{dref} for additional performance assessment. A 200-A step shift in the reference active current i_{dref} is expected at $t = 0.1$ s. The simulation result i_d and i_{dref} is shown in Figure 3.29(a), which suggests the tracking of the desired active current reached quickly. Additionally, simulations for changes in filter parameter L have been done to illustrate the resilience of the suggested AIT2FAC. Figure 3.30 illustrates i_d and i_q with a 10% change in L as a simulation result. The collected findings show that the suggested AIT2FAC is robust when L varies.

The tracking performance of the proposed AIT2FAC is compared with the type 1 fuzzy controller disclosed in [70] to assess its efficiency compared with other controllers. Figure 3.31 depicts the suggested AIT2FAC and type1 fuzzy controller's tracking performance, demonstrating the used AIT2FAC's superior control performance.

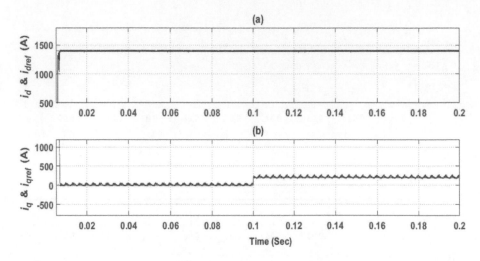

FIGURE 3.26 (a) i_d and i_{dref}; (b) i_q and i_{qref}. Power factor changes tracking.

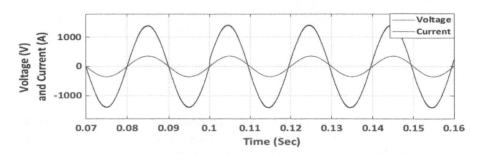

FIGURE 3.27 Voltage and current waveforms. Power factor changes tracking.

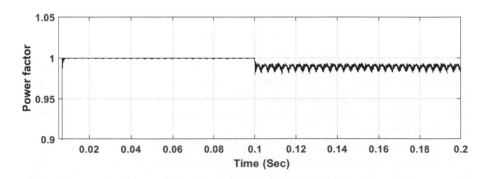

FIGURE 3.28 Power factor changes tracking.

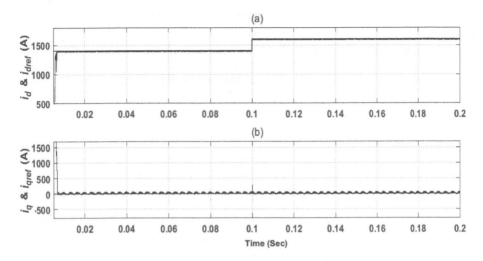

FIGURE 3.29   (a) i_d and i_{dref}; (b) i_q and i_{qref}; (c) voltage and current waveforms.

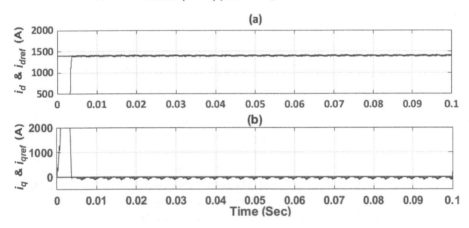

FIGURE 3.30   (a) i_d and i_{dref}; (b) i_q and i_{qref}; (c) voltage and current waveforms, with 10% increase in L.

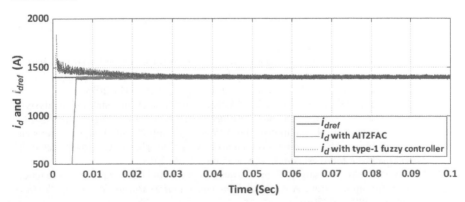

FIGURE 3.31   Performance of the proposed AIT2FAC and type 1 fuzzy controller.

TABLE 3.6

Performance Measures

Controller	Settling Time (sec)	Max Overshoot (%)	IAE	ITAE
Type 1 fuzzy controller	0.015	14.29	14.8505	1.8575
AIT2FAC	0.006	0.0	9.0070	0.9931

For the case of the unity power factor, Table 3.6 shows the tracking performance metrics for the type 1 fuzzy controller and proposed AIT2FAC

When it comes to tracking, the desired reference, Table 3.6 shows that the suggested controller surpasses the type 1 fuzzy controller. Furthermore, the suggested AIT2FAC has much lower integral absolute error (IAE) and integral time of absolute error (ITAE) performance indexes, according to the data. As a consequence, in terms of tracking performance, the proposed AIT2FAC surpasses the type 1 fuzzy controller.

3.3.6 CONCLUSION

This work provides adaptive interval type 2 fuzzy approximation control to solve the uncertainty and nonlinearity concerns of PV GCIS. Without the previous knowledge of the controlled system mathematical model, the suggested controller uses a fuzzy system estimate and the feedback linearization concept to estimate the controlled system's unknown nonlinear functions and monitor given reference values. The great capability of IT2FLS to deal with uncertainty is used. The updated law of the control parameters and the closed-loop stability were established using the Lyapunov function. The suggested controller was tested for different operations, and the results revealed that the suggested AIT2FAC is beneficial and has good tracking performance. Furthermore, in terms of multiple tracking performance parameters, the suggested AIT2FAC surpasses the type 1 fuzzy controller (ITAE).

REFERENCES

[1] A. Zahedi, "A review of drivers, benefits, and challenges in integrating renewable energy sources into electricity grid," *Renewable and Sustainable Energy Reviews*, vol. 15, no. 9, pp. 4775–4779, 2011.

[2] T. Dragičević, X. Lu, J. C. Vasquez, and J. M. Guerrero, "DC microgrids—Part II: A review of power architectures, applications, and standardization issues," *IEEE Transactions on Power Electronics*, vol. 31, no. 5, pp. 3528–3549, 2015.

[3] A. K. Abbasi and M. W. Mustafa, "Mathematical model and stability analysis of inverter-based distributed generator," *Mathematical Problems in Engineering*, 2013.

[4] S. Hadisupadmo, A. N. Hadiputro, and A. Widyotriatmo, "A small signal state space model of inverter-based microgrid control on single phase AC power network," *Internetworking Indonesia Journal*, vol. 8, pp. 71–76, 2016.

[5] P. Monica and M. Kowsalya, "Control strategies of parallel operated inverters in renewable energy application: A review," *Renewable and Sustainable Energy Reviews*, vol. 65, pp. 885–901, 2016.

[6] M. Castilla, J. Miret, A. Camacho, J. Matas, and L. G. de Vicuña, "Reduction of current harmonic distortion in three-phase grid-connected photovoltaic inverters via resonant current control," *IEEE Transactions on Industrial Electronics*, vol. 60, no. 4, pp. 1464–1472, 2011.

[7] G. Shen, X. Zhu, J. Zhang, and D. Xu, "A new feedback method for PR current control of LCL-filter-based grid-connected inverter," *IEEE Transactions on Industrial Electronics*, vol. 57, no. 6, pp. 2033–2041, 2010.

[8] F. Huerta, D. Pizarro, S. Cobreces, F. J. Rodriguez, C. Giron, and A. Rodriguez, "LQG servo controller for the current control of *LCL* grid-connected voltage-source converters," *IEEE Transactions on Industrial Electronics*, vol. 59, no. 11, pp. 4272–4284, 2011.

[9] A. Merabet, L. Labib, A. M. Ghias, C. Ghenai, and T. Salameh, "Robust feedback linearizing control with sliding mode compensation for a grid-connected photovoltaic inverter system under unbalanced grid voltages," *IEEE Journal of Photovoltaics*, vol. 7, no. 3, pp. 828–838, 2017.

[10] M. Mahmud, M. Hossain, H. Pota, and N. Roy, "Robust nonlinear controller design for three-phase grid-connected photovoltaic systems under structured uncertainties," *IEEE Transactions on Power Delivery*, vol. 29, no. 3, pp. 1221–1230, 2014.

[11] S. Mohiuddin, M. A. Mahmud, A. Haruni, and H. Pota, "Design and implementation of partial feedback linearizing controller for grid-connected fuel cell systems," *International Journal of Electrical Power & Energy Systems*, vol. 93, pp. 414–425, 2017.

[12] X. Zhang, Y. Wang, C. Yu, L. Guo, and R. Cao, "Hysteresis model predictive control for high-power grid-connected inverters with output LCL filter," *IEEE Transactions on Industrial Electronics*, vol. 63, no. 1, pp. 246–256, 2015.

[13] F.-R. López-Estrada, D. Rotondo, and G. Valencia-Palomo, "A review of convex approaches for control, observation and safety of linear parameter varying and Takagi-Sugeno systems," *Processes*, vol. 7, no. 11, pp. 814, 2019.

[14] T. Hornik and Q.-C. Zhong, "A current-control strategy for voltage-source inverters in microgrids based on H_∞ and repetitive control," *IEEE Transactions on Power Electronics*, vol. 26, no. 3, pp. 943–952, 2010.

[15] H. M. Hasanien, "An adaptive control strategy for low voltage ride through capability enhancement of grid-connected photovoltaic power plants," *IEEE Transactions on Power Systems*, vol. 31, no. 4, pp. 3230–3237, 2015.

[16] S. G. Jorge, C. A. Busada, and J. A. Solsona, "Frequency-adaptive current controller for three-phase grid-connected converters," *IEEE Transactions on Industrial Electronics*, vol. 60, no. 10, pp. 4169–4177, 2012.

[17] X. Li, H. Zhang, M. B. Shadmand, and R. S. Balog, "Model predictive control of a voltage-source inverter with seamless transition between islanded and grid-connected operations," *IEEE Transactions on Industrial Electronics*, vol. 64, no. 10, pp. 7906–7918, 2017.

[18] R. Errouissi, S. Muyeen, A. Al-Durra, and S. Leng, "Experimental validation of a robust continuous nonlinear model predictive control based grid-interlinked photovoltaic inverter," *IEEE Transactions on Industrial Electronics*, vol. 63, no. 7, pp. 4495–4505, 2015.

[19] E. Harirchian and T. Lahmer, "Improved rapid visual earthquake hazard safety evaluation of existing buildings using a type-2 fuzzy logic model," *Applied Sciences*, vol. 10, no. 7, p. 2375, 2020.

[20] Q. A. Tarbosh *et al.*, "Review and investigation of simplified rules fuzzy logic speed controller of high performance induction motor drives," *IEEE Access*, vol. 8, pp. 49377–49394, 2020.

[21] M. Pushpavalli and N. Jothi Swaroopan, "KY converter with fuzzy logic controller for hybrid renewable photovoltaic/wind power system," *Transactions on Emerging Telecommunications Technologies*, p. e3989, 2020.

[22] S. Mumtaz, S. Ahmad, L. Khan, S. Ali, T. Kamal, and S. Z. Hassan, "Adaptive feedback linearization based neurofuzzy maximum power point tracking for a photovoltaic system," *Energies*, vol. 11, no. 3, p. 606, 2018.

[23] M. Hosseinzadeh and F. R. Salmasi, "Power management of an isolated hybrid AC/DC micro-grid with fuzzy control of battery banks," *IET Renewable Power Generation*, vol. 9, no. 5, pp. 484–493, 2015.

[24] E. Harirchian and T. Lahmer, "Developing a hierarchical type-2 fuzzy logic model to improve rapid evaluation of earthquake hazard safety of existing buildings," in *Structures*, 2020, vol. 28: Elsevier, pp. 1384–1399.

[25] K. Mittal, A. Jain, K. S. Vaisla, O. Castillo, and J. Kacprzyk, "A comprehensive review on type 2 fuzzy logic applications: Past, present and future," *Engineering Applications of Artificial Intelligence*, vol. 95, p. 103916, 2020.

[26] G. Chen and T. T. Pham, *Introduction to Fuzzy Sets, Fuzzy Logic, and Fuzzy Control Systems*. CRC Press, 2000.

[27] M. Hannan, Z. A. Ghani, A. Mohamed, and M. N. Uddin, "Real-time testing of a fuzzy-logic-controller-based grid-connected photovoltaic inverter system," *IEEE Transactions on Industry Applications*, vol. 51, no. 6, pp. 4775–4784, 2015.

[28] S. Muyeen and A. Al-Durra, "Modeling and control strategies of fuzzy logic controlled inverter system for grid interconnected variable speed wind generator," *IEEE Systems Journal*, vol. 7, no. 4, pp. 817–824, 2013.

[29] P.-Z. Lin, C.-F. Hsu, and T.-T. Lee, "Type-2 fuzzy logic controller design for buck DC-DC converters," in *The 14th IEEE International Conference on Fuzzy Systems, 2005. FUZZ'05*, 2005: IEEE, pp. 365–370.

[30] N. Altin, "Interval type-2 fuzzy logic controller based maximum power point tracking in photovoltaic systems," *Advances in Electrical and Computer Engineering*, vol. 13, no. 3, pp. 65–71, 2013.

[31] A. H. El Khateb, N. Rahim, and J. Selvaraj, "Type-2 fuzzy logic approach of a maximum power point tracking employing sepic converter for photovoltaic system," *Journal of Clean Energy Technologies*, vol. 1, no. 1, pp. 41–44, 2013.

[32] N. Altin, "Single phase grid interactive PV system with MPPT capability based on type-2 fuzzy logic systems," in *2012 International Conference on Renewable Energy Research and Applications (ICRERA)*, 2012: IEEE, pp. 1–6.

[33] M. A. Hannan, Z. A. Ghani, M. M. Hoque, P. J. Ker, A. Hussain, and A. Mohamed, "Fuzzy logic inverter controller in photovoltaic applications: Issues and recommendations," *IEEE Access*, vol. 7, pp. 24934–24955, 2019.

[34] F.-J. Lin, K.-C. Lu, T.-H. Ke, B.-H. Yang, and Y.-R. Chang, "Reactive power control of three-phase grid-connected PV system during grid faults using Takagi–Sugeno–Kang probabilistic fuzzy neural network control," *IEEE Transactions on Industrial Electronics*, vol. 62, no. 9, pp. 5516–5528, 2015.

[35] F.-J. Lin, K.-C. Lu, and B.-H. Yang, "Recurrent fuzzy cerebellar model articulation neural network based power control of a single-stage three-phase grid-connected photovoltaic system during grid faults," *IEEE Transactions on Industrial Electronics*, vol. 64, no. 2, pp. 1258–1268, 2016.

[36] L.-X. Wang, *A Course in Fuzzy Systems and Control*. Prentice-Hall, Inc., 1996.

[37] L. X. Wang and H. Ying, "Adaptive fuzzy systems and control: design and stability analysis," *Journal of Intelligent and Fuzzy Systems-Applications in Engineering and Technology*, vol. 3, no. 2, p. 187, 1995.

[38] H. A. Yousef and M. A. Wahba, "Adaptive fuzzy mimo control of induction motors," *Expert Systems with Applications*, vol. 36, no. 3, pp. 4171–4175, 2009.

[39] H. M. Nguyen, *"Advanced control strategies for wind energy conversion systems,"* Idaho State University, 2013.

[40] C. Zhou *et al.*, "An improved direct adaptive fuzzy controller of an uncertain PMSM for web-based e-service systems," *IEEE Transactions on Fuzzy Systems*, vol. 23, no. 1, pp. 58–71, 2014.

[41] X. Liu, D. Zhai, J. Dong, and Q. Zhang, "Adaptive fault-tolerant control with prescribed performance for switched nonlinear pure-feedback systems," *Journal of the Franklin Institute*, vol. 355, no. 1, pp. 273–290, 2018.

[42] Y.-X. Li and G.-H. Yang, "Adaptive fuzzy fault tolerant tracking control for a class of uncertain switched nonlinear systems with output constraints," *Journal of the Franklin Institute*, vol. 353, no. 13, pp. 2999–3020, 2016.

[43] A. Yazdani *et al.*, "Modeling guidelines and a benchmark for power system simulation studies of three-phase single-stage photovoltaic systems," *IEEE Transactions on Power Delivery*, vol. 26, no. 2, pp. 1247–1264, 2010.

[44] P.-C. Chen, P.-Y. Chen, Y.-H. Liu, J.-H. Chen, and Y.-F. Luo, "A comparative study on maximum power point tracking techniques for photovoltaic generation systems operating under fast changing environments," *Solar Energy*, vol. 119, pp. 261–276, 2015.

[45] D. Lalili, A. Mellit, N. Lourci, B. Medjahed, and C. Boubakir, "State feedback control and variable step size MPPT algorithm of three-level grid-connected photovoltaic inverter," *Solar Energy*, vol. 98, pp. 561–571, 2013.

[46] H. K. Khalil and J. W. Grizzle, *Nonlinear Systems*. Prentice Hall: Upper Saddle River, NJ, 2002.

[47] T. J. Ross, *Fuzzy Logic with Engineering Applications*. John Wiley & Sons, 2005.

[48] Matlab M. MATLAB R2018b. The MathWorks: Natick, MA, USA. 2018.

[49] A. Arzani, P. Arunagirinathan, and G. K. Venayagamoorthy, "Development of optimal PI controllers for a grid-Tied photovoltaic inverter," in *2015 IEEE Symposium Series on Computational Intelligence*, 2015: IEEE, pp. 1272–1279.

[50] Y.-J. Liu, W. Wang, S.-C. Tong, and Y.-S. Liu, "Robust adaptive tracking control for nonlinear systems based on bounds of fuzzy approximation parameters," *IEEE Transactions on Systems, Man, and Cybernetics-Part A: Systems and Humans*, vol. 40, no. 1, pp. 170–184, 2009.

[51] Y. Li, S. Tong, Y. Liu, and T. Li, "Adaptive fuzzy robust output feedback control of nonlinear systems with unknown dead zones based on a small-gain approach," *IEEE Transactions on Fuzzy Systems*, vol. 22, no. 1, pp. 164–176, 2013.

[52] Z. Zhang and J. Dong, "A novel H∞ control for T–S fuzzy systems with membership functions online optimization learning," *IEEE Transactions on Fuzzy Systems*, vol. 30, no. 4, pp. 1129–1138, 2021.

[53] S. Hou and J. Fei, "A self-organizing global sliding mode control and its application to active power filter," *IEEE Transactions on Power Electronics*, vol. 35, no. 7, pp. 7640–7652, 2019.

[54] L. A. Zadeh, "The concept of a linguistic variable and its application to approximate reasoning—I," *Information Sciences*, vol. 8, no. 3, pp. 199–249, 1975.

[55] N. N. Karnik and J. M. Mendel, "Introduction to type-2 fuzzy logic systems," in *1998 IEEE International Conference on Fuzzy Systems Proceedings. IEEE World Congress on Computational Intelligence (Cat. No. 98CH36228)*, 1998, vol. 2: IEEE, pp. 915–920.

[56] N. N. Karnik, J. M. Mendel, and Q. Liang, "Type-2 fuzzy logic systems," *IEEE Transactions on Fuzzy Systems*, vol. 7, no. 6, pp. 643–658, 1999.

[57] Q. Liang and J. M. Mendel, "Interval type-2 fuzzy logic systems: theory and design," *IEEE Transactions on Fuzzy systems*, vol. 8, no. 5, pp. 535–550, 2000.

[58] J. R. Nayak, B. Shaw, and B. K. Sahu, "Application of adaptive-SOS (ASOS) algorithm based interval type-2 fuzzy-PID controller with derivative filter for automatic generation control of an interconnected power system," *Engineering Science and Technology, an International Journal*, vol. 21, no. 3, pp. 465–485, 2018.

[59] J. C. Cortes-Rios, E. Gómez-Ramírez, H. A. Ortiz-de-la-Vega, O. Castillo, and P. Melin, "Optimal design of interval type 2 fuzzy controllers based on a simple tuning algorithm," *Applied Soft Computing*, vol. 23, pp. 270–285, 2014.

[60] Z. Zhang and J. Dong, "Observer-based interval Type-2 L_2 – $L\infty/H\infty$ mixed fuzzy control for uncertain nonlinear systems under measurement outliers," *IEEE Transactions on Systems, Man, and Cybernetics: Systems*, vol. 51, no. 12, 7652–7662, 2020.

[61] P. Melin, "Interval type-2 fuzzy logic applications in image processing and pattern recognition," in *2010 IEEE International Conference on Granular Computing*, 2010: IEEE, pp. 728–731.

[62] P. Sharma and P. Bajaj, "Accuracy comparison of vehicle classification system using interval type-2 fuzzy inference system," in *2010 3rd International Conference on Emerging Trends in Engineering and Technology*, 2010: IEEE, pp. 85–90.

[63] J. S. Martínez et al., "Experimental validation of a type-2 fuzzy logic controller for energy management in hybrid electrical vehicles," *Engineering Applications of Artificial Intelligence*, vol. 26, no. 7, pp. 1772–1779, 2013.

[64] T.-C. Lin, H.-L. Liu, and M.-J. Kuo, "Direct adaptive interval type-2 fuzzy control of multivariable nonlinear systems," *Engineering Applications of Artificial Intelligence*, vol. 22, no. 3, pp. 420–430, 2009.

[65] F. Baghbani, M.-R. Akbarzadeh-T, and A. Akbarzadeh, "Indirect adaptive robust mixed H2/H∞ general type-2 fuzzy control of uncertain nonlinear systems," *Applied Soft Computing*, vol. 72, pp. 392–418, 2018.

[66] S. Hou, Y. Chu, and J. Fei, "Adaptive type-2 fuzzy neural network inherited terminal sliding mode control for power quality improvement," *IEEE Transactions on Industrial Informatics*, vol. 17, no. 11, 7564–7574, 2021.

[67] Z. Zhang and J. Dong, "Fault-tolerant containment control for IT2 fuzzy networked multiagent systems against denial-of-service attacks and actuator faults," *IEEE Transactions on Systems, Man, and Cybernetics: Systems*, vol. 52, no. 4, 2213–2224, 2021.

[68] H. Acikgoz, R. Coteli, B. Dandil, and F. Ata, "Experimental evaluation of dynamic performance of three-phase AC–DC PWM rectifier with PD-type-2 fuzzy neural network controller," *IET Power Electronics*, vol. 12, no. 4, pp. 693–702, 2019.

[69] H. Acikgoz, "Real-time adaptive speed control of vector-controlled induction motor drive based on online-trained Type-2 Fuzzy Neural Network Controller," *International Transactions on Electrical Energy Systems*, vol. 30, no. 12, pp. e12678, 2020.

[70] M. Shadoul, H. Yousef, R. Al Abri, and A. Al Hinai, "Adaptive fuzzy control of three-phase grid-connected inverter," in *2021 12th International Renewable Engineering Conference (IREC)*, 2021: IEEE, pp. 1–6.

[71] M. Shadoul, H. Yousef, R. A. Abri, and A. Al-Hinai, "Adaptive fuzzy approximation control of PV grid-connected inverters," *Energies*, vol. 14, no. 4, pp. 942, 2021.

[72] J. M. Mendel, R. I. John, and F. Liu, "Interval type-2 fuzzy logic systems made simple," *IEEE Transactions on Fuzzy Systems*, vol. 14, no. 6, pp. 808–821, 2006.

[73] M. Hosseinzadeh and M. J. Yazdanpanah, "Performance enhanced model reference adaptive control through switching non-quadratic Lyapunov functions," *Systems & Control Letters*, vol. 76, pp. 47–55, 2015.

[74] S. Sastry and M. Bodson, *Adaptive Control: Stability, Convergence, and Robustness*. Prentice-Hall, 1989.

4 Partial Shading Detection Method in Photovoltaic Systems

Waleed Al-Abri, Hassan Yousef, Rashid Al-Abri, Amer Al-Hinai

CONTENTS

4.1 INTRODUCTION

Several parameters can affect the performance of solar photovoltaic (PV) systems, including solar irradiance, ambient temperature, partial shading conditions (PSCs), cloud cover, and other causes. It is thought that PSCs are primarily responsible for reducing the life span and power output of solar panels [1]. Furthermore, PSCs are one of the main contributors to power mismatches within PV arrays or modules [2]. Also, the conventional maximum power point tracking (MPPT) systems, such as perturb and observation (P&O) [3–7], constant voltage technique (CVT) [8], short circuit technique (SCT) [9], open circuit voltage technique (OVT) [10], and incremental conductance (IC) [11–13], fail to maximize the PV system output power during PSC [1]. The development of intelligent MPPT methods for improving conventional MPPT systems has been the focus of many research studies in the last several years. Such methods include fuzzy logic (FL) [14–16], support vector machines (SVMs) [17, 18], neural networks [19, 20], Grey Wolf optimization (GWO) [21], particle swarm optimization (PSO) [22, 23], firefly algorithm (FA) [24], and simulated annealing (SA) [25]. The intelligent MPPT method seems to be quicker, more effective, and more stable than conventional methods [26]. Nevertheless, this method is complicated and difficult to implement [26] and experiences a large amount of calculation burden [27]. Also, the intelligent MPPT methods are unable to guarantee the

DOI: 10.1201/9781003307433-4

convergence of global MPP operations across all PSCs. Therefore, both conventional MPPTs and intelligent MPPTs should have a mechanism to determine the global MPP under PSCs, such as I–V curve tracer, electronics [28, 29], resistance [30], and capacitance [31, 32], or a soft I–V curve sweeper, such as ant colony optimization (ACO) [33], Jaya-differential evolution (DE) [34], cuckooo search (CS) [35], DE [36], chaotic search [37], and artificial bee colony (ABC) [21]. If these searching techniques are not properly triggered in response to PSCs, the PV system's power output will be further reduced. It is therefore very important to have methods of detecting PSCs that can accurately detect when they occur.

Several partial shading detection (PSD) methods were presented in [38–43]. PSD methods rely on the monitoring of changes between successive measurements of PV array output parameters, such as I_{pv_array}, V_{pv_array}, and P_{pv_array}. According to [38–43], it may be possible to detect PSCs by comparing the rate of change in PV array output parameters with threshold values. For example, [41, 42] considered 10% to 20% of P_{mpp} and 5% of the nominal power, respectively, to detect PSCs. For the same rate change, [43] set the threshold at 15%. According to [38–40, 44], to avoid incorrectly detecting PSCs even at low solar irradiance levels, the authors emphasized the need to normalize the change rate in the output parameters of the solar array. As an example, based on [38], to detect PSCs, it is necessary to normalize the change in power rate to the rate change of voltage as expressed in Equation (4.1).

$$\text{Partial shading index (PSI)} = \frac{\Delta P}{\Delta V.P}\bigg|V_{mpp\_array} \tag{4.1}$$

Another suggestion came from [40, 44], who stated that comparing the true power to the maximum power calculated by Equation (4.2) is sufficient to detect PSCs. They claim that if the ΔP is higher than 0.1, it is considered that PSC takes place. Research for [45] followed the same approach as that for [40], but applied a smaller statistical threshold (0.05).

$$\Delta P = \left|\frac{P_{in} - P_{MPP}}{P_{MPP}}\right| \tag{4.2}$$

Reference [46] proposes an accurate method to detect PSCs. This method involves placing a voltage sensor on each PV module terminal. A PSC is alerted when the panel voltage goes down more than 1 volt. In contrast to [46, 47], only one voltage sensor was used. A controller and switching matrix were used by the author to measure the voltage across PV panels. A block diagram for the method is shown in Figure 4.1. However, these methods are expensive methods and are unpractical for large arrays.

4.2 THE PROPOSED PARTIAL SHADING DETECTION METHOD

4.2.1 OVERVIEW

PV arrays change their voltage output with variations in their output current, which are directly influenced by the solar radiation they receive. According to Figure 4.2, the PV array's voltage rises in the power region as the array current decreases.

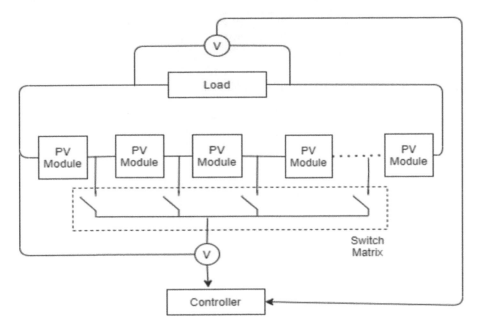

FIGURE 4.1   Schematic for the presented partial shading detection (PSD) method stated in [47].

PSCs may decrease or increase the array current depending on their type. PSCs, such as object shading and row-to-row shading, can fully block sunlight, which, in turn, causes conduction in bypassing diodes, resulting in a drop in the array voltage. Alternatively, some PSC conditions (e.g., clouds) result in a reduction of sunlight incident on the panels. Due to this reduction in output current, array voltage increases. The array voltage changes are therefore an indication of all PSC types.

A PSD method is proposed in this chapter based on the previous observation that tracks successive voltage fluctuations across a PV array. The flowchart in Figure 4.3 illustrates how the proposed method works. The proposed method counts the number of times the difference between the previous and the next voltage is consecutively positive or negative. It declares the occurrence of PSCs whenever successive voltage changes show the same sign for a certain number of consecutive changes

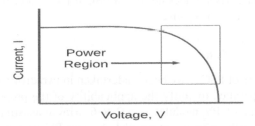

FIGURE 4.2   I–V characteristics curve of a photovoltaic panel/array.

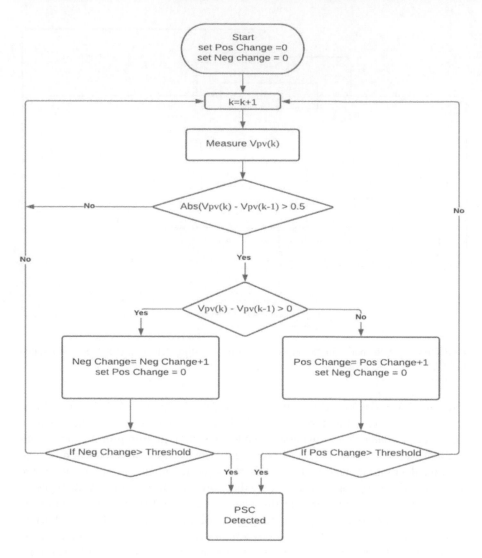

FIGURE 4.3 Proposed method flowchart.

(the threshold value). Also, it does not flag a PSC if the sign of the voltage changes repeatedly from positive to negative.

4.2.2 Experimental Setup

Physical simulations of PSCs have been undertaken in two different PV array types: series and series parallel, to verify the applicability of the proposed method. The series-parallel array of PV modules is a 2 × 6 array consisting of two modules arranged in series, with six strings in parallel. A series PV array is made up of 15 PV panels. The PV arrays with series and parallel connections have MPP voltages of

TABLE 4.1

System Parameters

Parameter	Series Array	Series-Parallel Array
Manufacturer	ZNShine Solar	BenQ Solar
Module	ZXH6-LD72	PM096B00
V_{MPP}	38.4 V	54.7 V
I_{MPP}	8.91 A	6.09 A
P_{MPP}	345 W	333 W
V_{OC}	46.9 V	64.9 V
I_{SC}	9.42 A	6.58 A
Number of PV cells	72	96
Bypassing diodes	3	3

576 and 109.4 V, respectively. In Table 4.1, the specifications of the PV modules used in a series PV array and in a series-parallel PV array are presented under standard test conditions (STCs). To measure the voltage during PSC simulations, a Fluke 435 Power Quality Meter is connected to the output terminals of the PV array. Figure 4.4 shows the experimental setup of both configurations of solar PV arrays.

4.2.3 EXPERIMENT PROCEDURE

To achieve the minimum amount of shading, a variety of shading scenarios were simulated, from shading single PV cells to shading entire PV panels. Simulations of two types of PSCs were conducted: object shading and cloud shading. Object shading is emulated using blinding material (see Figure 4.5b) and cloud shading is emulated using tinting material (see Figure 4.5a). The Fluke 435 Power Quality Meter

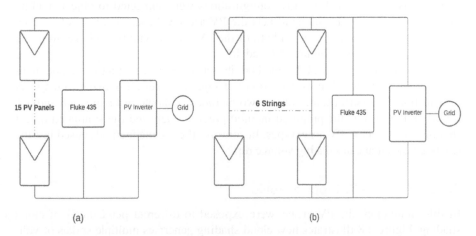

FIGURE 4.4 Experimental setup of (a) a series PV array and (b) a series-parallel PV array.

(a)   (b)

FIGURE 4.5   (a) Simulation of cloud shading using a window film. (b) Simulation of object shading by putting a thick paper sheet.

was used to record the voltage of the PV array every 0.25 s during simulations. The experiment took place during a clear day with solar light intensity between 350 and 750W/m² and an ambient temperature of 36°C.

4.3   RESULTS AND DISCUSSION

The figures in this section show the PV array voltage waveform in blue and the detection signal output of the proposed PSD method in red. The output signal of the proposed PSD is either one (indicating that a PSC has been discovered) or zero (indicating that PSC does not exist).

4.3.1   SCENARIO I: SHADING OF OBJECT

During this study, both PV array configurations were subjected to object shading to see how this type of shading affects the PV array voltage. This type of shading appears to have similar effects on both of the PV arrays, with the voltage dropping rapidly as can be seen in Figures 4.6 and 4.7.

To evaluate the efficacy of this method, its act is compared with the method mentioned in [40, 44], which uses the ratio of change in power normalized to the level of detection. The performance of the two methods is shown in Figures 4.6 and 4.7. Based on the results, the proposed method precisely detected the simulated object shading cases in both PV array types. In contrast, the second method sensed just the shading cases that caused a big voltage dip.

4.3.2   SCENARIO II: CLOUD SHADING

In this simulation, the PV arrays were exposed to different percentages of cloud shading. Figure 4.8 illustrates how cloud shading generates multiple spikes of voltage in series-parallel PV arrays under simulation.

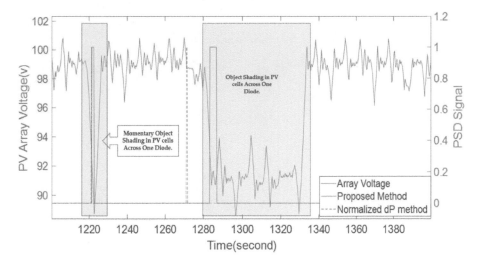

FIGURE 4.6 Voltage waveforms of the series-parallel PV array and the PSD output signals during the emulation of object shading.

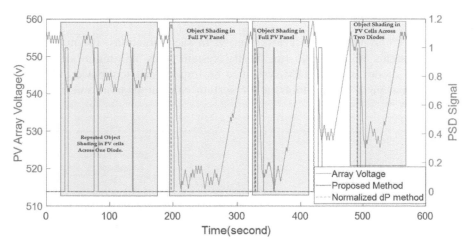

FIGURE 4.7 Voltage waveforms of the series PV array and the PSD output signals during the emulation of object shading.

In contrast to the series-parallel PV array, the series PV array shows a rise in voltage during simulation as seen in Figure 4.9. This is due to the shaded PV panels bottlenecking the entire array current during the shading period, which in turn causes the array voltage to increase.

The PSD output signal presented by a red line, as shown in Figures 4.8 and 4.9, illustrates the method's ability to detect this type of shading rapidly and exactly in both PV array types. On the other hand, the detection methods in [40, 44] could not able to sense the emulated cloud shading cases in both PV array types.

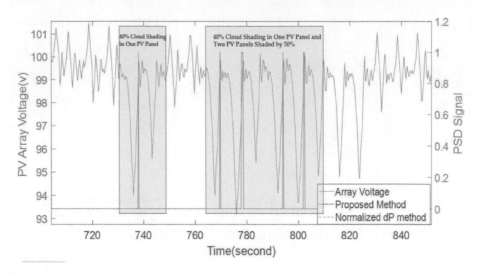

FIGURE 4.8 Array voltage waveform of the series-parallel PV and PSD signal during the emulation of cloud shading cases.

4.4 CONCLUSION

In PSCs, PV systems produce less energy. In such situations, the majority of MPPT systems are unable to run the PV system at its maximum because the bypassing diodes cause the PV system to be trapped at a low power point. By using global peak finding tools, a PV system can output more energy at the

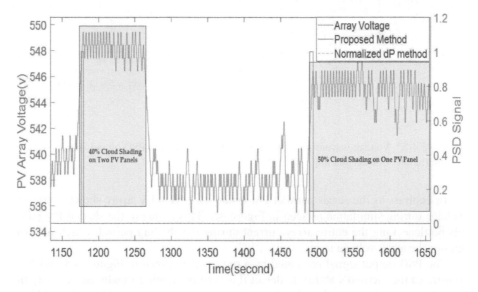

FIGURE 4.9 Array voltage waveform of the series PV and PSD signal during the emulation of the cloud shading cases.

global MPP. These tools, still, drop the output of PV systems, as they force the PV system to run outside of its power region for scanning the I–V curve. Thus, global peak searching should only be implemented during PSCs. This chapter presented an accurate method for sensing PSCs using the voltage sign. In the PSD method, a PSC is predicted whenever the sign of successive voltage changes persist a certain number of times. The effectiveness of the proposed method was tested by simulating a number of cloud and object shading patterns on-site using two types of PV array (series and series parallel). Results indicate that the proposed method correctly identified all simulated shading patterns. Further, the performance of this method is superior to a detection method utilizing a normalized measurement of power change.

REFERENCES

1. da Luz, C.M.A., E.M. Vicente, and F.L. Tofoli, *Experimental evaluation of global maximum power point techniques under partial shading conditions*. Solar Energy, 2020. **196**: p. 49–73.
2. Bai, J., et al., *Characteristic output of PV systems under partial shading or mismatch conditions*. Solar Energy, 2015. **112**: p. 41–54.
3. Abdel-Salam, M., M. El-Mohandes, and M. El-Ghazaly, *An efficient tracking of MPP in PV systems using a newly-formulated P&O-MPPT method under varying irradiation levels*. Journal of Electrical Engineering & Technology, 2019. **15**: p. 501–513.
4. Alik, R. and A. Jusoh, *An enhanced P&O checking algorithm MPPT for high tracking efficiency of partially shaded PV module*. Solar Energy, 2018. **163**: p. 570–580.
5. Al-Majidi, S.D., M.F. Abbod, and H.S. Al-Raweshidy, *A novel maximum power point tracking technique based on fuzzy logic for photovoltaic systems*. International Journal of Hydrogen Energy, 2018. **43**(31): p. 14158–14171.
6. Manickam, C., et al., *Fireworks Enriched P&O algorithm for GMPPT and detection of partial shading in PV systems*. IEEE Transactions on Power Electronics, 2017. **32**(6): p. 4432–4443.
7. Ghamrawi, A., J.-P. Gaubert, and D. Mehdi, *A new dual-mode maximum power point tracking algorithm based on the Perturb and Observe algorithm used on solar energy system*. Solar Energy, 2018. **174**: p. 508–514.
8. Lasheen, M., et al., *Adaptive reference voltage-based MPPT technique for PV applications*. IET Renewable Power Generation, 2017. **11**(5): p. 715–722.
9. Sher, H.A., et al., *A single-stage stand-alone photovoltaic energy system with high tracking efficiency*. IEEE Transactions on Sustainable Energy, 2017. **8**(2): p. 755–762.
10. Veerapen, S., W. Huiqing, and D. Yang. Design of a novel MPPT algorithm based on the two stage searching method for PV systems under partial shading, in *2017 IEEE 3rd International Future Energy Electronics Conference and ECCE Asia (IFEEC 2017 - ECCE Asia)*. 2017.
11. Kumar, R., et al., *Global maximum power point tracking using variable sampling time and p-v curve region shifting technique along with incremental conductance for partially shaded photovoltaic systems*. Solar Energy, 2019. **189**: p. 151–178.
12. Motahhir, S., et al., *Modeling of photovoltaic system with modified incremental conductance algorithm for fast changes of irradiance*. International Journal of Photoenergy, 2018. **2018**: p. 1–13.
13. Yatimi, H., Y. Ouberri, and E. Aroudam, *Enhancement of power production of an autonomous PV system based on robust MPPT technique*. Procedia Manufacturing, 2019. **32**: p. 397–404.

14. Farajdadian, S. and S.M.H. Hosseini, *Design of an optimal fuzzy controller to obtain maximum power in solar power generation system.* Solar Energy, 2019. **182**: p. 161–178.
15. Li, X., et al., *Comprehensive studies on operational principles for maximum power point tracking in photovoltaic systems.* IEEE Access, 2019. **7**: p. 121407–121420.
16. Yilmaz, U., A. Kircay, and S. Borekci, *PV system fuzzy logic MPPT method and PI control as a charge controller.* Renewable and Sustainable Energy Reviews, 2018. **81**: p. 994–1001.
17. Kihal, A., et al., *An improved MPPT scheme employing adaptive integral derivative sliding mode control for photovoltaic systems under fast irradiation changes.* ISA Transactions, 2019. **87**: p. 297–306.
18. Pahari, O.P. and B. Subudhi, *Integral sliding mode-improved adaptive MPPT control scheme for suppressing grid current harmonics for PV system.* IET Renewable Power Generation, 2018. **12**(16): p. 1904–1914.
19. Hamdi, H., C. Ben Regaya, and A. Zaafouri, *Real-time study of a photovoltaic system with boost converter using the PSO-RBF neural network algorithms in a MyRio controller.* Solar Energy, 2019. **183**: p. 1–16.
20. Issaadi, S., W. Issaadi, and A. Khireddine, *New intelligent control strategy by robust neural network algorithm for real time detection of an optimized maximum power tracking control in photovoltaic systems.* Energy, 2019. **187**: p. 115881.
21. Sundareswaran, K., et al., *Enhanced energy output from a PV system under partial shaded conditions through artificial bee colony.* IEEE Transactions on Sustainable Energy, 2014. **6**(1): p. 198–209.
22. Ishaque, K. and Z. Salam, *A deterministic particle swarm optimization maximum power point tracker for photovoltaic system under partial shading condition.* IEEE Transactions on Industrial Electronics, 2012. **60**(8): p. 3195–3206.
23. Ishaque, K., et al., *An improved particle swarm optimization (PSO)–based MPPT for PV with reduced steady-state oscillation.* IEEE Transactions on Power Electronics, 2012. **27**(8): p. 3627–3638.
24. Sundareswaran, K., S. Peddapati, and S. Palani, *MPPT of PV systems under partial shaded conditions through a colony of flashing fireflies.* IEEE Transactions on Energy Conversion, 2014. **29**(2): p. 463–472.
25. Lyden, S. and M.E. Haque, *A simulated annealing global maximum power point tracking approach for PV modules under partial shading conditions.* IEEE Transactions on Power Electronics, 2015. **31**(6): p. 4171–4181.
26. Belhachat, F. and C. Larbes, *A review of global maximum power point tracking techniques of photovoltaic system under partial shading conditions.* Renewable and Sustainable Energy Reviews, 2018. **92**: p. 513–553.
27. Liu, Y.-H., J.-H. Chen, and J.-W. Huang, *A review of maximum power point tracking techniques for use in partially shaded conditions.* Renewable and Sustainable Energy Reviews, 2015. **41**: p. 436–453.
28. Campos, R.E., et al. Experimental analysis of a developed I-V curve tracer under partially shading conditions, in *2019 IEEE PES Innovative Smart Grid Technologies Conference - Latin America (ISGT Latin America).* 2019.
29. Ahmad, R., et al., *An MPPT technique for unshaded/shaded photovoltaic array based on transient evolution of series capacitor.* Solar Energy, 2017. **157**: p. 377–389.
30. van Dyk, E.E., A.R. Gxasheka, and E.L. Meyer, *Monitoring current–voltage characteristics and energy output of silicon photovoltaic modules.* Renewable Energy, 2005. **30**(3): p. 399–411.
31. Mahmoud, M.M., *Transient analysis of a PV power generator charging a capacitor for measurement of the I–V characteristics.* Renewable Energy, 2006. **31**(13): p. 2198–2206.
32. Spertino, F., et al., *Capacitor charging method for I–V curve tracer and MPPT in photovoltaic systems.* Solar Energy, 2015. **119**: p. 461–473.

33. Sundareswaran, K., et al., *Development of an improved P&O algorithm assisted through a colony of foraging ants for MPPT in PV system.* IEEE Transactions on Industrial Informatics, 2015. **12**(1): p. 187–200.

34. Kumar, N., et al., *Rapid MPPT for uniformly and partial shaded PV system by using JayaDE algorithm in highly fluctuating atmospheric conditions.* IEEE Transactions on Industrial Informatics, 2017. **13**(5): p. 2406–2416.

35. Ahmed, J. and Z. Salam, *A maximum power point tracking (MPPT) for PV system using Cuckoo Search with partial shading capability.* Applied Energy, 2014. **119**: p. 118–130.

36. Tajuddin, M.F.N., et al., *Evolutionary based maximum power point tracking technique using differential evolution algorithm.* Energy and Buildings, 2013. **67**: p. 245–252.

37. Zhou, L., et al., *New approach for MPPT control of photovoltaic system with muta-tive-scale dual-carrier chaotic search.* IEEE Transactions on Power Electronics, 2010. **26**(4): p. 1038–1048.

38. Ghasemi, M.A., H.M. Foroushani, and M. Parniani, *Partial shading detection and smooth maximum power point tracking of PV arrays under PSC.* IEEE Transactions on Power Electronics, 2015. **31**(9): p. 6281–6292.

39. Pillai, D.S., et al., *An accurate, shade detection-based hybrid maximum power point tracking approach for PV systems.* IEEE Transactions on Power Electronics, 2020. **35**(6): p. 6594–6608.

40. Wang, Y., Y. Li, and X. Ruan, *High-accuracy and fast-speed MPPT methods for PV string under partially shaded conditions.* IEEE Transactions on Industrial Electronics, 2015. **63**(1): p. 235–245.

41. Rizzo, S.A. and G. Scelba, *ANN based MPPT method for rapidly variable shading conditions.* Applied Energy, 2015. **145**: p. 124–132.

42. Ramyar, A., H. Iman-Eini, and S. Farhangi, *Global maximum power point tracking method for photovoltaic arrays under partial shading conditions.* IEEE Transactions on Industrial Electronics, 2016. **64**(4): p. 2855–2864.

43. Kim, R.-Y. and J.-H. Kim, *An improved global maximum power point tracking scheme under partial shading conditions.* Journal of International Conference on Electrical Machines and Systems, 2013. **2**(1): p. 65–68.

44. Wellawatta, T.R. and S.J. Choi, *Adaptive partial shading determinant algorithm for solar array systems.* Journal of Power Electronics, 2019. **19**(6): p. 1566–1574.

45. Al-Ramaden, A. and I.A. Smadi, Partial shading detection and global MPPT algorithm for PV system, in *2019 IEEE Jordan International Joint Conference on Electrical Engineering and Information Technology, JEEIT 2019 - Proceedings.* 2019.

46. Zbeeb, A., V. Devabhaktuni, and A. Sebak, Improved photovoltaic MPPT algorithm adapted for unstable atmospheric conditions and partial shading, in *2009 International Conference on Clean Electrical Power.* 2009.

47. Ma, J., et al., Automatic shading detection system for photovoltaic strings, in *Proceedings - International SoC Design Conference 2018, ISOCC 2018.* 2019.

5 Renewable Energy Management Concept

Amer Al-Hinai, Ahmed Al Maashri,
Rashid Al-Abri, Saira Al-Zadjali, Mana Al-Shekili

CONTENTS

5.1 THE ESSENCE OF DISPATCHABLE HYBRID POWER PLANT

The energy production from wind turbines and Photovoltaic (PV) systems is not reliable by itself as wind speed is intermittent and not all wind speeds can be harvested. Similarly, the solar irradiation is usually unpredictable, PV system output is variable during the day, and not available at night [1]. Variability and ramp events in output power are the key challenges to the system operators due to their impact on system balancing, reserves management, scheduling, and commitment of generation units [2]. As a result, wind and PV resources are nondispatchable. Integrating wind and PV systems economically into a conventional power grid necessitates the elimination of the complications caused by the intermittent behavior of these resources.

Then comes the need to aggregate wind turbines with the PV system output and integrate the Energy Storage System (ESS) to waive the intermittency of wind-PV and balance power generation. The ESS can store extra produced energy and use it

DOI: 10.1201/9781003307433-5

to compensate for the reduction in power generation and balance out the fluctuations of the wind-PV power in short and long periods.

In theory, the output power from Renewable Energy Sources (RESs) such as PV and wind turbines can be varied from zero to maximum available power. Nevertheless, maximum available power is variable as renewable sources (i.e., wind and solar irradiance) are variable throughout the day.

In [3], an Energy Management System (EMS) is introduced for a dispatchable hybrid renewable generation power plant. Using the synergy of wind and solar generation optimizes power output from wind, solar, and the Battery Energy Storage System (BESS) to maintain the delivered power from the power plant constant for a given amount of time. The proposed EMS is displayed in Figure 5.1.

The four main components of the proposed EMS are forecasting, aggregating, scheduling, and controlling, in which the forecaster predicts the wind speed and direction in addition to solar irradiance and temperature through meteorological data. Next, the forecasted data along with structure of the RESs are used to predict the available output power. The scheduler optimizes the system operation (i.e., maximize output power while minimizing fluctuation and use of BESS) by calculating set points of wind, PV, and storage systems. Finally, the control unit would execute these set points and supervise and control the output power of different system components [2].

In Figure 5.2, the forecasted power output from the hybrid power plant is divided into dispatchable and nondispatchable power. The dispatchable power is further divided into dispatched power (i.e., contracted by the system operator) and uncommitted power. The dispatched power is never higher than the dispatchable power and computed by the optimal scheduling unit to bid in the electricity spot market. The uncommitted power is the difference between the dispatched and dispatchable power. The nondispatchable power is the difference between the actual available power and the dispatchable power [3].

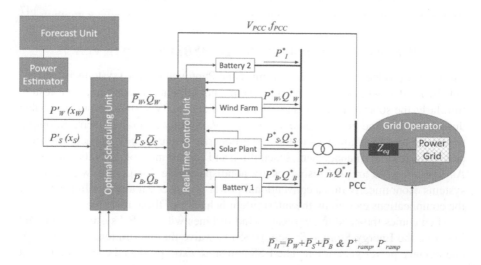

FIGURE 5.1 Proposed dispatchable hybrid power plant EMS [4].

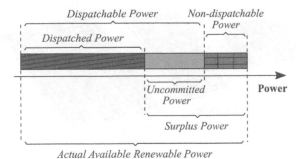

FIGURE 5.2   Organization of available power [5].

5.1.1 Photovoltaic (PV) Energy

It is expected, by 2040, that the output of PV systems will reach 7200 TWh according to the world energy outlook [6] exceeding the energy generated by hydropower. In terms of new installations, solar PV is considered the biggest installed capacity for power generation among the various types of RESs [7]. This is mainly due to the decreased cost of production and recent rapid advancement in the manufacturing process.

The building blocks of a PV array are solar cells connected in series and parallel. A solar cell is an electronic component that converts sun irradiance into electricity. Practically, a solar cell is manufactured from a mixture of semiconductor materials [i.e., commonly, Silicon (Si) and Germanium (Ge)] to form a p-n junction [5]. When light strikes this semiconductor material, the photon energy is absorbed; hence, a current and voltage are produced to develop electrical power.

The process of generating electrical current in a solar cell is called "light-generated current" and is accomplished through two stages. First, the absorption of photons creates electron-hole pairs. The second stage involves the collection of carriers by the p-n junction (i.e., holes are carriers in this case). This process stops the recombination by using a p-n junction to spatially separate the electron and the hole. An external circuit is utilized to facilitate the flow of charges, hence, a current is produced [5]. On the other hand, a voltage is developed in the solar cell by a process called "photovoltaic effect". A collection of light-generated carriers increases the number of electrons and holes in the n-type and p-type side in the p-n junction, respectively. Therefore, the electric field in the junction is reduced, leading to an increased diffusion current. An equilibrium state is reached where voltage is developed across the p-n junction [8].

Solar cells are connected in series and parallel to form a PV module. Typically, a module will have 35 or 72 cells connected in series. PV modules are then arranged in series and parallel to make PV arrays as demonstrated in Figure 5.3. This configuration of solar cells and modules enables the development of large current and voltage at the terminals of PV arrays.

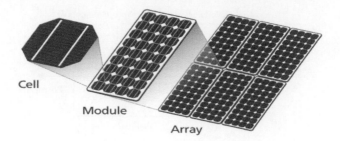

FIGURE 5.3   PV cell, module, and array [9].

5.1.1.1   PV Mathematical Model

A solar cell is ideally modeled as a current source in parallel with a single diode as illustrated in Figure 5.4. Nevertheless, to precisely represent a solar cell's nonlinear current-voltage characteristics, additional series and parallel resistors are added to the model.

From [10, 11], a current-voltage relationship can be derived as follows:

$$I = I_{ph} - I_d - I_{sh} \tag{5.1}$$

$$I = I_{ph} - I_o \left(\exp\left(\frac{q(V + R_s I)}{nkT} \right) - 1 \right) - \left(\frac{V + R_s I}{R_{sh}} \right) \tag{5.2}$$

where I_{ph} is the photocurrent produced by the incidence of sunlight on the solar cell, I_o is the diode reverse saturation current, V is the output voltage, and R_s and R_{sh} are the series and shunt resistance, respectively. The diode thermal voltage V_t is a constant that is represented by $V_t = \frac{KT}{q}$, where T is the temperature in Kelvin, k is the Boltzmann constant (1.380649×10^{-23} J/K), and q is the electron charge ($1.60217663 \times 10^{-19}$ C). The diode ideality factor n is introduced in the dominator that ranges between 1 and 2 to represent how closely the diode matches the ideal diode [11, 12].

Because the shunt resistance R_{sh} value is very large, its effect can be neglected to further simplify the model; introducing a four-parameter model and Equation (5.2) becomes:

$$I = I_{ph} - I_o \left(\exp\left(\frac{q(V + R_s I)}{nkT} \right) - 1 \right) \tag{5.3}$$

Both four-parameter and five-parameter models are available in real-time digital simulator (RTDS).

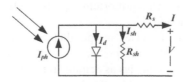

FIGURE 5.4   A single diode showing a five-parameter model of a solar cell.

5.1.1.2 PV Maximum Power Point Tracking (MPPT)

Multiple various techniques are presented in the literature such as perturb and observe (P&O), incremental conductance, and ripple correlation. The most common technique is the P&O [13], where the PV reference voltage is varied using a boost converter whose duty cycle is controlled by the Maximum Power Point Tracking (MPPT) controller. The resulting current and voltage are then sampled and compared with the previous power value [13]. The PV I–V characteristics curve in Figure 5.5 is utilized for this method. The curve clearly shows that there exists a voltage value that corresponds to the maximum power point. The P&O algorithms are demonstrated in Figure 5.6.

The RTDS PV model is integrated with two optional analytical methods for approximating the PV maximum power point. The first method is the fractional open circuit voltage where the voltage at maximum point $V_{mp} = (80\%)V_{oc}$, depending on the voltage temperature coefficient [11].

The second method depends on the Lambert function approximation [14], where N_s is the number of series-connected modules in a PV array, whereas N_{cs} is the number of series-connected cells in a module. The maximum power point is given by

$$V_{mp} = V_{im} - \left(I_{im} - \frac{V_{im}}{R_{sh}} \right) R_s \tag{5.4}$$

where:

$$V_{im} = N_s N_{cs} n V_t \left(Lambert \left(\frac{I_{ph} \exp(1)}{I_o} \right) - 1 \right) \tag{5.5}$$

$$I_{im} = \frac{V I_o}{N_s N_{cs} n V_t} \exp \left(\frac{V}{N_s N_{cs} n V_t} \right) \tag{5.6}$$

5.1.2 Wind Energy

The wind is a bulky movement of air in a certain direction and with a specific speed. It flows on a large-scale area causing wind turbines to rotate. This rotation of the

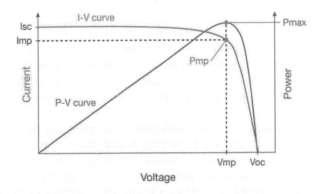

FIGURE 5.5 PV I–V and P-V characteristics curve.

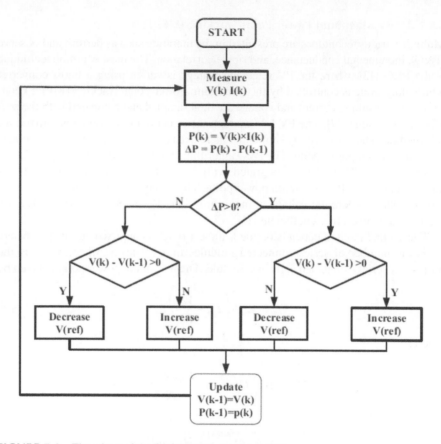

FIGURE 5.6   Flowchart of the P&O MPPT algorithm [15].

wind turbines generates wind power. Moreover, this transformation from the wind speed to wind power is presented in the following equation:

$$P = \frac{1}{2} pAv^3 \tag{5.7}$$

where $P(W)$ is the output wind power from the wind farms (WFs). It depends on the air density p (kg/m³), the area of the wind turbine's rotor $A(m^2)$, and the wind speed $v(m/s)$. It is clear from Equation (5.7) that the relation between the output power and the wind speed is a nonlinear cubic relationship. Therefore, any slight change in the wind speed will reflect into a significant error in the estimated power outcome from the farm.

Wind energy depends on wind speed, and wind speed can be definitely influenced by land surface, obstacles, turbine height, and other environmental conducts. In addition, unpredictability and the variability characteristics of the wind speed are the fundamental worries for the power system operators. Therefore, wind power generation will be impacted too, and as a result, the reliability of the wind power will be unsatisfactory for providing steady electricity into power systems. To solve these issues, the power generation systems must be scheduled ahead to get information about the wind speed variations. One of the most popular methods used in this case is the wind speed prediction, which will help power the generation system to be prepared ahead for the estimated upcoming wind power.

To predict wind speed, various methods have been developed and are divided into different categories in the literature. The methods are divided into three basic categories: persistence, physical, and statistical.

5.1.2.1 Persistence Method

The persistence method uses a simple assumption that an event like wind speed at a certain future time will be the same as of now. In other words, the prediction after some time steps ahead will bear the same value of the first forecast [33]. For example, if the measured wind speed at a time (i) is $v(i)$, then the forecasted wind speed at a time $(i+1)$ will be the same $v(i)$. Mathematically, this can be presented as follows:

$$v(i+1) = v(i) \tag{5.8}$$

For the wind speed prediction, the persistence method is useful and shows accurate forecasting results for a very short time. However, the accuracy of the persistence method gradually decreases as the time steps are increased [34]. Besides simplicity, the persistent method is found to be economical as well. Hence, the literature includes this method as a benchmark to compare and validate the forecasting model's improvement over the persistence method [35].

5.1.2.2 Physical Method

The physical method is a mathematical model that uses physical considerations such as land surfaces, obstacles, temperature, pressure measurements, etc., to make the best prediction for a certain event. One of the most popular examples of the physical prediction method is numeric weather prediction (NWP), where the complex mathematical approaches are operated using weather data such as land surface roughness, obstacles, and other metrological variables [35].

NWPs are used in long-term predictions, as it is difficult to get information in a short time by these methods because they require a high computing machine to run the complex mathematical models, as well as their concomitant high cost. These facts limit the usefulness of NWP forecasts for long-term prediction (i.e., > 6 hours ahead) [36, 37].

Landberg et al. [38] used NWP, which performed better in forecasting the wind speed after 6 hours than the persistence method. Moreover, the models that are not using NWP have quite good accuracy for the first few steps of the prediction, in the short term, but they are generally useless for long-horizon predictions [39]. El-Fouly et al. [40] took into consideration many physical aspects such as surface roughness, obstacles, orography, etc. The model was forecasting from 3 to 24 hours and it performed better than the persistence method in terms of the correlation coefficient between the output and the target values. The correlation coefficient for the proposed model was 87% at predictions 3 hours ahead, which then decreased as the predicted horizon steps were increased. Finally, 58% of the correlation was obtained at predictions that were 24 hours ahead. In addition, in terms of the average percentage error, the proposed model performed well in obtaining an error of 9.04% for predictions that were 3 hours ahead and 28.23% for predictions that were 24 hours ahead, which were better than the persistence method.

For short-term forecasting, Negnevitsky et al. [41] tuned NWP and applied digital elevation models (DEMs) and model output statistics (MOS) with it. This method showed some realization; however, it has been proved that NWPs are not satisfactorily suitable for short-term forecasting. This concept again was used by Negnevitsky and Potter [42], and they illustrated that to achieve better accuracy in short-term forecasting, NWP needs to be employed with MOS and DEM. Similarly, Watson et al. [43] employed NWP/MOS to forecast the wind speed and the direction of the wind for the UK grid system. The predictions were based on an hourly time step of up to 18 hours. The authors concluded that the NWP/MOS method can save 15% to 25% of the fossil fuel cost with its accurate prediction.

5.1.2.3 Statistical Method

Statistical models are time series representations that depend on historical data to estimate the patterns by adjusting the parameter of the model based on the difference between the actual value and the predicted one. Then, these patterns are employed to make the mathematical models to be used in predictions. In time series applications, these models can be divided into two methods: conventional time series and artificial neural network (ANN). These models are good for short-term predictions and are inexpensive as they do not require high-performance computing (HPC) to run the model [35]. Moreover, Neural network methods surpassed these conventional methods, because of their fast computing power, excellent ability in learning from previous data, and in getting minimum mean-square errors between the predicted values and actual ones [6]. Essentially, the accuracy of these models depends on the amount of training data and the model structure, whereas the latter is used to train the data.

5.2 THE RATIONALE FOR SIZING HYBRID RENEWABLE RESOURCES

As mentioned in the introduction, wind-PV energy systems will contribute a substantial amount of the globally produced energy in the upcoming years, so a strategy is needed for controlling their power output, especially while using Hybrid Wind-Solar Photovoltaic Power Systems (HWSPSs), which have quite a variable and uncertain power output. Energy storage has been used to smooth the power output from wind and solar sources, but significant challenges remain [27]. Studies have shown that the reliability of a hybrid wind-PV system is twice that of either technology used alone [28]. This is attributed to the inherent complementary characteristics of wind and solar energy. Solar has several characteristics that are quite different from wind, such as self-correlation and strong production in the middle of the day.

The integration of renewable resources into power generation, and of wind energy in particular, has become the primary focus for investments in the power generation sector. WFs play a significant role in satisfying the demand for energy, but constructing a WF involves technical challenges and cost implications. Wind profile is crucial for defining a turbine's capacity for energy production. For instance, the cubical relationship between wind speed and wind output power means that any deviation in wind speed has a severe impact on the turbine's output power and, consequently, on the system's performance [29]. Therefore, a hierarchy involving the turbine's specification, the wind profile, and the geography of the WF site is required to build a farm that utilizes resources efficiently and performs well. One of the main challenges in WF design is the allocation of wind turbines with the aim of decreasing the inconsistency in output power [9].

The main steps involved in the design of a WF are as follows: site identification, technical and economic analysis, regulatory compliance, micro-siting, and construction [30]. The WF layout optimization (WFLO) problem is complex because many factors have to be considered. The wake effect on the farm is one of the main factors affecting the WF output power. The wake effect is the change in the effective speed of the wind that powers downstream turbines as a result of the turbulence caused by upstream turbines. Optimization techniques must be used to obtain an optimal layout. The WFLO problem has been tackled in the literature with different fitness functions and optimization techniques. Heuristic techniques are more effective than analytical optimization methods, even when optimizing a simple objective such as the farm's output power [15].

According to the literature, the genetic algorithm (GA) is one of the heuristic methods commonly used to solve similar mixed-integer problems. This is due to its ability to avoid local optima and its high likelihood of locating global optima [31]. The authors in [9] used the GA to find the optimal number and location of wind turbines for optimal power production.

The wake effect has been introduced into the WFLO problem using different models in discrete or continuous search spaces. These include Gaussian [32, 33], Jensen's [34, 35], Larsen's [36], and Frandsen's [37] models. These models differ in their level of complexity and, therefore, in their computational burden. An early attempt to use Jensen's model for WFLO was carried out by Mosetti et al. [9, 38] in 1993, employing the GA to minimize the cost per unit energy produced. They investigated the problem for different wind scenarios with varying complexities. The wind scenarios commonly evaluated in the literature include wind with a constant speed and direction, wind with a constant speed but variable direction, and wind with a variable speed and variable direction. Grady et al. [39] improved the WFLO using the same objective function, the same wake-effect model, and the same grid strategy as Mosetti et al. [38]. The improvement in results is attributed to the use of different GA parameters. Most of the published research work [40–42] on WF optimization uses Jensen's wake-effect model that assumes a uniform velocity inside the wake cone.

A study [43] used Jensen's wake-effect model and optimizes the layout using the binary real coded GA (BRCGA) based on a local search approach. The underlying principle is to find the turbine location with the power output of each turbine. The GA solution has been improved using the local search technique to find the optimal solution near the solution found by the BRCGA. The researchers tested their method on two wind scenarios: wind with multiple directions and multiple speeds and wind with multiple directions but constant speed. They demonstrated that using the BRCGA yields results closer to the results achieved by [39], but incurs a higher computational burden. Particle swarm optimization (PSO) was used in [6] to solve a multi-objective WFLO problem by maximizing the power output and minimizing the costs. The optimization formulation constraints are the dimensions of the land lot and a clearance distance of eight times the rotor diameter between any two wind turbines, following the industry standard.

Heuristic optimization techniques are most commonly used to tackle WFLO problems [33, 44]. In [45], an approach was developed for optimizing the layout of a large WF, whereas the approach in [46] is applicable for a WF with few wind turbines. Other techniques, such as quadratic integer programming, multipopulation GA, and nonlinear programming, were used in [47, 48], and [49], respectively. A heuristic ant colony optimization algorithm is presented in [50] for the layout optimization of a WF with eight turbines and a cost-integrated model. Amaral et al. [51] used the GA and PSO for offshore WFLO and compared the performance of both approaches. They concluded that PSO performs faster, whereas the GA yields better results. A comprehensive review is presented in [52] for the GA, PSO, and fuzzy methods. These heuristic techniques are widely used in the optimal design of hybrid RESs. The algorithm and mechanism of the GA are explained in detail in [53]. The GA is widely used for solving the WFLO problem; it was used in 75% of the WFLO studies [9]. Thus, the literature review shows that to solve this nonlinear, constrained optimization problem, most of the previous studies used heuristic algorithms.

Cabling work is a significant part of the total cost of WF construction. Cabling work includes small cabling between wind turbines and the large collector cable that connects the wind turbines to the point of common coupling (PCC). During WF construction, cabling between the wind turbines is one of the debatable design tasks. As stated in [54, 55], the installation and capital cost of the cabling work significantly contributes to the total cost of the project. In addition, power losses constitute an operation cost, which affects the long-term financial profitability of the project. Hence, cable work should not be neglected. Simple techniques such as geometric programming were used in [56] to obtain the optimal electrical layout. The first attempts at WF cabling adopted a greedy algorithm [57], which starts by adding a turbine to a location that provides the maximum power output. This process repeated for all the turbines while considering the wake effect of the existing turbines. The process ends when the maximum number of turbines is reached [58]. This algorithm connects each turbine to its closest neighbor, which is not necessarily the optimal connection. In other words, the solution provided by the greedy algorithm is far from optimal, because each step depends on the formation created in the preceding step and lacks a holistic view of the land layout. Hence, in this study, the cabling works are integrated with the WF layout using heuristic optimization techniques to build the integrated tool.

Smoothing of wind power output is also necessary to keep the output power more reliable. Power smoothing methods can be divided into methods based on energy storage (indirect methods) and methods not based on energy storage (direct methods) [59]. Indirect methods use ESSs; the BESS is often used due to its fast dynamic response [60]. The BESS is normally used to compensate for the required supply or to absorb power to make the system stable and to ensure that it is within operational limits [61].

In some studies, mathematical smoothing techniques were used to generate a reference power line. Some of these techniques are moving average (MAV) [62], double MAV [63], moving median, exponential moving average (EMA) [64], and wavelet decomposition [65]. The smoothness is adjusted by manipulating the window size, as in [66], whereas double MAV is used for PV output power. In addition, the EMA of wind power was used as a reference output power in [64]. Moreover, the authors of [67] used the MAV to mitigate fluctuations in solar PV output. In [68], the averaged output powers of PV and wind were used as a reference to determine the BESS capacity. In [63], the BESS capacity was defined based on the MAV of PV output power. The authors of [69] used two predetermined BESS capacities, one designed for energy, and the other for power smoothing, depending on the reference PV power produced from MAV and a low-pass filter. A hierarchal MAV was also used in [70] to get the optimal BESS use to mitigate the fluctuations in PV output power. The studies show that MAV is a simple way to filter out power fluctuations and generate a reference power line.

According to the literature, MAV is widely used to smooth noisy signals. For smoothing, the BESS is operated to make up the difference between the actual signal and the MAV signal. But MAV depends on the past time series data, which are different from the current value of the fluctuating variable. The main problem with MAV is the memory effect feature, which means that the approach depends on the data from the past [71]. This means that the BESS is used more than necessary. This shortens the BESS' life span. In addition, the memory effect can also cause oversmoothing, which increases the required BESS capacity. To avoid the memory effect, other smoothing techniques are used in this study, like locally weighted linear regression (LWLR), Gaussian, and Savitzky-Golay (SG).

The increasing level of wind and PV plant use has given rise to serious problems resulting from the intermittent and stochastic nature of these resources. This chapter proposes a hierarchal tool for developing an optimal HWSPS by selecting the suitable smoothing tool. Furthermore, actual recorded wind speeds and wind directions are used to mimic the real scenario.

In addition, compliance with the system regulations is important to ensure that the system is reliable and to avoid penalties for violating regulations.

5.3   THE RATIONALE FOR FORECASTING IN POWER SYSTEMS APPLICATIONS

The importance of accurate wind speed and solar irradiance forecasts to power systems operations cannot be overemphasized. Wind turbine power curves are highly nonlinear and the cubic relationship between wind speed and wind power means that a small error in wind speed prediction corresponds to a very large error in predicted power output. The main use of forecasting renewable power output (solar and wind) in power systems is balancing the network. The low-capacity factors and output variability of these plants means that the transmission system operator must grapple with ensuring that enough reserves are available to account for shortfalls in

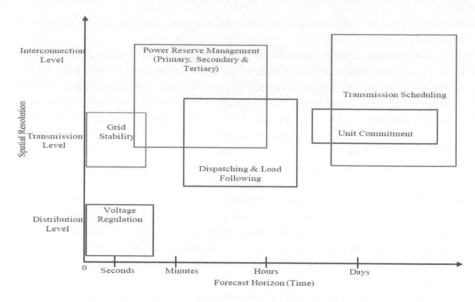

FIGURE 5.7 Applications of forecasts with respect to spatial resolution and temporal horizon.

wind and or PV power production. Other factors such as forecast of congestion, network losses, electricity market participation, etc., are also significant [72]. Accurate forecasts result in enormous technical and economic benefits. In [73], it is reported that the Western Electricity Coordinating Council (WECC) system would save up to $28 million annually for a 10% reduction in wind speed forecasting errors at 14% renewable power penetration. These savings significantly rise with increasing levels of renewable energy penetration. In summary, the forecasts aid the power system operators in preparing the system for oncoming high ramp rates of renewable generators for optimal dispatching to ensure supply reliability, plan for ancillary services market to mitigate against intermittent generation, enable dispatching of quick-start generators in advance, and ensuring optimal operation scheduling to enable dispatch of available renewable generation. Figure 5.7, which is a modification from [74], shows various applications of forecasts with respect to their spatial resolution and temporal horizon.

5.4 THE RATIONALE FOR DISPATCHING IN POWER SYSTEMS APPLICATIONS

Dispatchability refers to the ability of the power system units (e.g., generator and demand) to change their power injection/draw within the announced operational limits, per the request of the system operator. Power system operation is centered around the key dispatchability feature of conventional generation units. Solar and wind power plants can potentially vary their output between zero and the maximum available power [75]. However, due to the intermittency of the available power (i.e., variable wind speed and solar irradiance), they are inherently nondispatchable, meaning there is no guarantee that the expected amount of power will be available on demand, for a given period of time. Power system penetration of renewables and in specific solar and wind generation has substantially increased worldwide. Realizing 100% renewables-based power systems is a hot topic under investigation [76]. Dispatchable renewable generation is the holy grail of 100% renewables-based power systems. It enables the participation of renewables in several electricity markets and potentially reduces the costs of reserve and flexibility associated with the integration of renewables. Dispatchable renewable generation entails adequately sized and controlled ESSs to compensate for power deficiency due to the variability of renewables [77]. Nguyen et al. [78] investigated BESS capacity and power rating with the aid of a rule-based dispatch strategy, using mean wind power as the reference dispatch power.

Abdullah et al. [79] proposed to initially carry out day-ahead scheduling of wind power using scenario-based stochastic programming to maximize the expected revenue. In the real-time control, battery storage units are provided with set points, individually, to maintain an even state of charge. The aim was to meet the committed power as well as prolong the BESS' lifetime. Due to the fast exhausting of BESS, Gholami et al. [60] proposed dividing it into two sections and controlling them to meet the mean wind power forecast, as a reference for dispatch. Hence, BESS lifetime can be increased by increasing the depth of charge/discharge in each section by only allowing charging or discharging in several consecutive periods. Instead of such a division, the authors of [80, 81] considered hybrid ESS (HESS) in their proposed design. Sequential control of hybrid flywheel-battery ESS to smooth wind power is presented in [80]. The reference power is optimized for minimum deviation from the forecast wind power. The share of each storage resource in the ESS is determined such that the cost of operation is minimized. To avoid inefficient use of the battery requires simultaneous charge/discharge of ESS resources and limiting the battery power rating with a state-of-charge-based logarithmic function.

In [81], a statistical approach was taken to size a supercapacitor–battery HESS for dispatchable wind power. The study proposed handling high-frequency variations of wind power with the supercapacitor and managed the power flow between the storage resources to maintain a safe battery power ramp rate.

5.5 THE RATIONALE FOR CONTROLLING IN POWER SYSTEM APPLICATIONS

Currently, the implemented control strategies of grid-connected renewable systems are designed to track the power output maximum point and export it to the grid [82–85]. Although this method reduces the complexity of attached control systems, it proposes challenges to the power grid operator. However, the grid operator is required to operate RESs as a base load and compensate for intermittencies by assigning conventional synchronous generators. Hence, limiting the penetration of renewable systems in power grids and increasing the cost of renewable systems integration.

Several strategies are proposed in the literature to minimize the renewable energy power output fluctuations by integrating an energy source of a controllable power output (i.e., mostly a system of a storage system).

In [67], a fuel cell is integrated with a PV system to form a hybrid system to minimize power fluctuations and handle the variations of a locally connected load. Nevertheless, this approach does not solve the problem of unpredicted output power. Hence, it excludes the mentioned hybrid system from the unit commitment planning. An effective power dispatch control strategy to improve generation schedulability and supply reliability of a WF is proposed in [86]. A BESS is integrated with the proposed system while taking into account the uncertainty in the wind generation and the energy prices. However, no actual power forecasting is considered, and the target is still to track the maximum power output. A hierarchical power sharing control in direct current (DC) microgrids is introduced and analyzed in [87]. The study redefines the design of renewable energy control schemes. It accounts for all the inherited challenges and complexity associated with hybrid renewable systems. Having said that, solving the intermittency challenge associated with renewable energy and transforming it into dispatchable energy necessitates the integration of a storage system, developing forecasting techniques, and implementing a hierarchical control system.

Fundamentally, a hybrid power plant consists of multiple different energy source types combined with an integrated storage system. Therefore, a suitable EMS and a control scheme must be implemented to deal with different energy sources individually as well as collectively. Achieving a properly designed EMS and control system requires knowledge about classifications of different energy units installed in the hybrid power plant. Different energy units can be classified by their operating mode, dispatchable or nondispatchable units. Furthermore, storage units must be identified along with their capacity and state of charge (SOC).

Based on the available different energy units and the control target to be achieved, all converters can be modeled as a voltage source or a current source. Hence, power flow control and voltage regulation require proper control strategy.

5.5.1 HIERARCHICAL POWER SHARING IN CONVENTIONAL POWER SYSTEMS

In conventional power systems, frequency and voltage levels must remain within acceptable limits to achieve a stable and reliable power system. Hence, all grid-connected generators have to participate in voltage and frequency regulation. In conventional power systems, reactive power is provided by synchronous generators and reactive power compensators. Active power is supplied by synchronous generators. However, the load of the grid is a stochastic variable and cannot be predicted accurately. Hence, some generators are reserved for adjusting the differences between the forecasted and actual load [82, 83]. This is achieved by implementing a droop controller in the governor system. A droop characteristics system facilitates the distribution of load variances between parallel generators. Because the frequency is the same all over the grid, different generators can be managed collaboratively using droop characteristics. Figure 5.8 illustrates the effects of primary and secondary controllers when utilizing the frequency droop characteristics. The response of two synchronous generators is depicted, where P_1 and P_2 refer to the power output of generator 1 and 2, respectively, while ω^* is the reference frequency and ω is operating frequency. The steady-state frequency drops at ω. The secondary controller compensates the difference in frequency and shifts up the droop characteristics

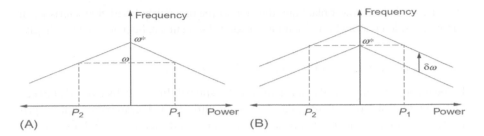

FIGURE 5.8 Droop characteristics of two synchronous generators. (a) Response of primary controller. (b) Response of secondary controller [91].

until operating frequency equals the required reference frequency. The reference frequency value is determined by a tertiary controller [88–90].

These fundamentals have been implemented in alternating current (AC) microgrids. The same can be applied to a hybrid power plant discussed in this chapter, because they both share similar characteristics, complexities, and challenges as discussed in the following section.

5.5.2 HYBRID POWER PLANT (HPP) AS A SYSTEM OF SYSTEMS (SoS)

It should be pointed out that from a control system design perspective, a hybrid power plant is considered a complex system consisting of various systems. Almost all those systems are nonlinear by definition and a strong cross-coupling exists between them. Identifying such a system as a system of systems (SoS) leads to designing an effective control structure to overcome the challenges presented by the hybrid power plant [92]. The concept of an SoS is now widespread and has entered several domains, including defense, IT, health care, manufacturing, energy, space stations, and exploration [93, 94].

An SoS is uniquely defined by specific characteristics mentioned in [77]. The SoS must show the following characteristics:

1. **Operational independence:** All the constituent systems within the SoS architecture operate independently and do not interfere with neighboring systems in their functionality.
2. **Managerial independence:** The constituent systems continue to operate on their own unperturbed by the SoS. In other words, they are responsible for their autonomous operation.
3. **Evolutionary development:** The SoS is not designed as a single unit, and is rather flexible, and can accommodate numerous new systems or eliminate systems that are no longer necessary.
4. **Emergent behavior:** All the constituent systems function as a collective unit to accomplish a common objective, which cannot be achieved by a single-component system.
5. **Geographic distribution:** The distribution of the subsystems is sequential to facilitate the flow of information among them.

Because the hybrid power plant qualifies for all the previously listed characteristics, it can be considered as an SoS. Therefore, a control system can be planned accordingly.

5.5.3 CONTROL HIERARCHY IN HYBRID POWER PLANT

In the theory of hierarchical control, different responsibilities are distributed between different control levels. Three main levels are built in the hierarchical control, primary controller, secondary controller, and tertiary controller. Figure 5.9 displays the structure of these levels along with their relationship to each other [92].

In a primary controller, a reliable operation is concerned in addition to regulating voltage and frequency according to the grid code. The output of active P and reactive Q power is ensured according to the reference value and limits allowed.

The secondary controller monitors the total power generated at the PCC and acts as a supervisory central controller to compensate for any deviations existing from the connected energy sources units. The secondary controller can receive power reference value either from a tertiary controller or the grid operator.

Finally, the tertiary controller makes decisions economically governing the flow of power from the hybrid power plant to the grid while complying with the rules of the electricity market. The approved results of the tertiary controller are communicated with the secondary controller as well as the grid operator.

5.5.4 INVERTERS CONTROLLER IN HIERARCHICAL CONTROL

Because the hybrid power plant is composed of RESs, those RESs are commonly interfaced with power electronic converters (i.e., DC-AC or AC-DC-AC). Power electronic inverters (i.e., mostly VSI) can be utilized to facilitate the implementation of hierarchical control. The VSI controls both the magnitude and the phase of the voltage output. Hence, it has the ability to control both active power P and reactive

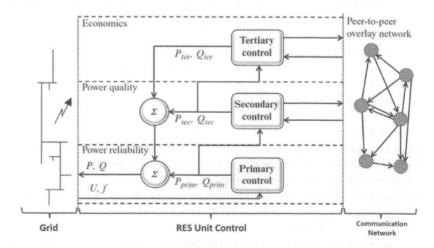

FIGURE 5.9 Hierarchical control of structure [92].

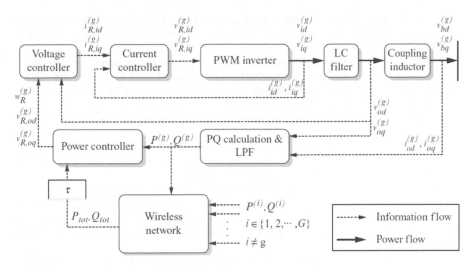

FIGURE 5.10   Common voltage source inverter control, with Low-Pass Filter (LPF) [92].

power Q. Figure 5.10 demonstrates the basic common blocks and control loops of VSI controllers in addition to their interactions. [92]-[118].

REFERENCES

[1] Al-Maamary, H.M., H.A. Kazem, and M.T. Chaichan, *The impact of oil price fluctuations on common renewable energies in GCC countries.* Renewable and Sustainable Energy Reviews, 2017. **75**: p. 989–1007.

[2] IRENA, REthinking Energy 2017: Accelerating the global energy transformation. Retrieved January 17, from http://www.irena.org/publications/2017/Jan/REthinking-Energy-2017-Accelerating-the-global-energy-transformation. 2017.

[3] REN21 Secretariat, P., Renewables 2017 Global Status Report. 2017.

[4] Bakhtvar, M. et al., *Dispatchable renewable generation: Oman's harweel wind farm potential.* Cigre GCC, 2019.

[5] Aggarwal, S. and M. Gupta, *Wind power forecasting: A review of statistical models-wind power forecasting.* International Journal of Energy Science, 2013. **3**(1).

[6] Lei, M., L. Shiyan, J. Chuanwen, L. Hongling, and Z. Yan, *A review on the forecasting of wind speed and generated power.* Renewable and Sustainable Energy Reviews, 2009. **13**: p. 915–920.

[7] Mathiesen, P., D. Rife, and C. Collier, Forecasting solar irradiance variability using the analog method. in Photovoltaic Specialists Conference (PVSC), IEEE 43rd. 2016.

[8] Lorenz, E., et al., *Irradiance forecasting for the power prediction of grid-connected photovoltaic systems.* IEEE Journal of Selected Topics in Applied Earth Observations and Remote Sensing, 2009. **2**(1): p. 2–10.

[9] Samlex, "The Difference Between Solar Cell, Module & Array I Samlex Solar." https://www.samlexsolar.com/learning-center/solar-cell-module-array.aspx (accessed Nov. 03, 2020).

[10] El-Fouly, T.H., E.F. El-Saadany, and M.M. Salama, *One day ahead prediction of wind speed and direction.* IEEE Transactions on Energy Conversion, 2008. **23**(1): p. 191–201.

[11] Tascikaraoglu, A., et al., *Compressive spatio-temporal forecasting of meteorological quantities and photovoltaic power.* IEEE Transactions on Sustainable Energy, 2016. **7**(3): p. 1295–1305.

[12] Kim, J.-G., et al., *Daily prediction of solar power generation based on weather forecast information in Korea.* IET Renewable Power Generation, 2017 **11**(10): p. 1268–1273.

[13] Bivona, S., G. Bonanno, R. Burlon, *Stochastic models for wind speed forecasting.* Energy Conversion and Management, 2011. **52**(2): p. 1157–1165.

[14] Mohandes, M.A., et al., *Support vector machines for wind speed prediction.* Renewable Energy, 2004. **29**(6): p. 939–947.

[15] Bai, W., M.R. Abedi, and K.Y. Lee, *Distributed generation system control strategies with PV and fuel cell in microgrid operation.* Control Engineering Practice, 2016. **53**: p. 184–193, doi: 10.1016/j.conengprac.2016.02.002.

[16] Wan, C., *et al.*, *Probabilistic forecasting of wind power generation using extreme learning machine.* IEEE Transactions on Power Systems, 2014. **29**(3): p. 1033–1044.

[17] Xu, Y., Q. Hu, and F. Li, *Probabilistic model of payment cost minimization considering wind power and its uncertainty.* IEEE Transactions on Sustainable Energy, 2013. **4**(3): p. 716–724.

[18] The Grid Code. *National Grid Electricity Transmission plc, London.* U.K., 2013, vol. 5, Revision 12.

[19] Pudjianto, D., C. Ramsay, and G. Strbac, *Virtual power plant and system integration of distributed energy resources.* IET Renewable Power Generation, 2007. **1**(1): p. 10–16.

[20] Albadi, M.H. and E.F. El-Saadany, *Overview of wind power intermittency impacts on power systems.* Electric Power Systems Research, 2010. **80**(6): p. 627–632.

[21] Albadi, M. and E. El-Saadany, *Comparative study on impacts of wind profiles on thermal units scheduling costs.* IET Renewable Power Generation, 2011. **5**: p. 26–35.

[22] Barklund, E., N. Pogaku, M. Prodanovic, C. Hernandez-Aramburo, and T. C. Green, *Energy management in autonomous microgrid using stability-constrained droop control of inverters.* IEEE Transactions on Power Electronics, 2008. **23**(5): p. 2346–2352.

[23] H. Nehrir et al., "A review of hybrid renewable/alternative energy systems for electric power generation: Configurations, control and applications," 2012 IEEE Power and Energy Society General Meeting, San Diego, CA, 2012, p. 1–1.

[24] Tani, A., M.B. Camara, and B. Dakyo, *Energy management in the decentralized generation systems based on renewable energy—ultracapacitors and battery to compensate the wind/load power fluctuations.* IEEE Transactions on Industry Applications, 2015. **51**(2), p. 1817–1827.

[25] Al-hinai A. et al., *Development of Energy Management System for Dispatchable Renewable Power Generation.* Project Proposal. Sultan Qaboos University.

[26] IEA, "Renewables – World Energy Outlook 2019 – Analysis - IEA," 2019. https://www.iea.org/reports/world-energy-outlook-2019/renewables#abstract (accessed Nov. 02, 2020).

[27] IEA, "World Energy Outlook 2019 – Analysis - IEA," 2019. https://www.iea.org/reports/world-energy-outlook-2019 (accessed Nov. 02, 2020).

[28] "PVEducation." https://www.pveducation.org/ (accessed Oct. 26, 2020).

[29] Palz, W., *Power for the World - The Emergence of Electricity from the Sun.* Belgium: Pan Stanford Publishing, 2010.

[30] RTDS, "PV ARRAY MODEL." RTDS Software Manuals, p. 1–16.

[31] Ropp, M.E. and S. Gonzalez, *Development of a MATLAB/Simulink Model of a single-phase grid-connected photovoltaic system.* IEEE Transactions on Energy Conversion, 2009. **24**(1): p. 195–202, doi: 10.1109/TEC.2008.2003206.

[32] Farivar, G., B. Asaei, and S. Mehrnami, *An analytical solution for tracking photovoltaic module MPP.* IEEE Journal of Photovoltaics, 2013. **3**(3): p. 1053–1061, doi: 10.1109/JPHOTOV.2013.2250332.

[33] Zhao, X., S. Wang, and T. Li, *Review of evaluation criteria and main methods of wind power forecasting.* Energy Procedia, 2011. **12**: p. 761–769.

[34] Wu, Y.-K. and J.-S. Hong, *A literature review of wind forecasting technology in the world.* in Power Tech, 2007 IEEE Lausanne, 2007: p. 504–509.

[35] Soman, S.S., H. Zareipour, O. Malik, and P. Mandal, "A review of wind power and wind speed forecasting methods with different time horizons," in *North American Power Symposium (NAPS)*, 2010: p. 1–8.

[36] Potter, C.W. and M. Negnevitsky, *Very short-term wind forecasting for Tasmanian power generation.* IEEE Transactions on Power Systems, 2006. **21**: p. 965–972.

[37] Candy, B., S.J. English, and S.J. Keogh, *A comparison of the impact of QuikScat and WindSat wind vector products on met office analyses and forecasts.* IEEE Transactions on Geoscience and Remote Sensing, 2009. **47**: p. 1632–1640.

[38] Landberg, L., *Short-term prediction of local wind conditions.* Journal of Wind Engineering and Industrial Aerodynamics, 2001. **89**: p. 235–245.

[39] Giebel, G., R. Brownsword, G. Kariniotakis, M. Denhard, and C. Draxl, *The state-of-the-art in short-term prediction of wind power: A literature overview.* Anemos Plus, 2011.

[40] El-Fouly, T., E. El-Saadany, and M. Salama, "One day ahead prediction of wind speed using annual trends," in *Power Engineering Society General Meeting IEEE.* 2006, p. 7.

[41] M. Negnevitsky, P. Johnson, and S. Santoso, "Short term wind power forecasting using hybrid intelligent systems," in *Power Engineering Society General Meeting.* IEEE 2007.

[42] M. Negnevitsky and C. W. Potter, "Innovative short-term wind generation prediction techniques," in *Power Systems Conference and Exposition. PSCE'06.* IEEE PES, 2006, p. 60–65.

[43] Watson, S., L. Landberg, and J. Halliday, *Application of wind speed forecasting to the integration of wind energy into a large scale power system.* IEE Proceedings-Generation, Transmission and Distribution, 1994. **141**: p. 357–362.

[44] Chompoo-Inwai, C., et al., *System impact study for the interconnection of wind generation and utility system.* IEEE Transactions on Industry Applications, 2005. **41**(1): p. 163–168.

[45] Liu, X., P. Wang, and P.C. Loh, A hybrid AC/DC microgrid and its coordination control. in IEEE Transactions on Smart Grid, 2011. **2**(2): p. 278–286.

[46] Albadi, M. and E. El-Saadany, *Overview of wind power intermittency impacts on power systems.* Electric Power Systems Research, 2010. **80**(6): p. 627–632.

[47] Shakoor, R., et al., *Wake effect modeling: A review of wind farm layout optimization using Jensen's model.* Renewable and Sustainable Energy Reviews, 2016. **58**: p. 1048–1059.

[48] Mathew, S., *Wind Energy: Fundamentals, Resource Analysis and Economics.* Springer, 2006, Vol. 1.

[49] González, J.S., et al., *A review and recent developments in the optimal wind-turbine micro-siting problem.* Renewable and Sustainable Energy Reviews, 2014. **30**: p. 133–144.

[50] Rao, S.S., *Engineering Optimization: Theory and Practice.* John Wiley & Sons, 2009.

[51] Carbajo Fuertes, F., C.D. Markfort, and F. Porté-Agel. *Incoming Wind and Wake Measurements of a Single 2.5 MW Wind Turbine Using Two Nacelle-Mounted Wind LiDARs for Analytical Wake Model Validation.* in *23rd Symposium on Boundary Layers and Turbulence.* 2018.

[52] Parada, L., et al., *Wind farm layout optimization using a Gaussian-based wake model.* Renewable Energy, 2017. **107**: p. 531–541.

[53] Politis, E.S., et al., *Modeling wake effects in large wind farms in complex terrain: the problem, the methods and the issues.* Wind Energy, 2012. **15**(1): p. 161–182.

[54] Jensen, N.O., *A Note on Wind Generator Interaction.* Riso National Laboratory, 1983.

[55] Larsen, G.C., *A Simple Wake Calculation Procedure.* Risø National Laboratory Risø-M-2760, 1988.

[56] Frandsen, S., *On the wind speed reduction in the center of large clusters of wind turbines.* Journal of Wind Engineering and Industrial Aerodynamics, 1992. **39**(1-3): p. 251–265.

[57] Mosetti, G., C. Poloni, and B. Diviacco, *Optimization of wind turbine positioning in large windfarms by means of a genetic algorithm.* Journal of Wind Engineering and Industrial Aerodynamics, 1994. **51**(1): p. 105–116.

[58] Grady, S., M. Hussaini, and M.M. Abdullah, *Placement of wind turbines using genetic algorithms.* Renewable Energy, 2005. **30**(2): p. 259–270.

[59] Wang, L., Y. Zhou, and J. Xu, *Optimal irregular wind farm design for continuous placement of wind turbines with a two-dimensional Jensen-Gaussian wake model.* Applied Sciences, 2018. **8**(12): p. 2660.

[60] Li, X., D. Hui, and X. Lai, *Battery energy storage station (BESS)-based smoothing control of photovoltaic (PV) and wind power generation fluctuations.* IEEE Transactions on Sustainable Energy, 2013. **4**(2): p. 464–473.

[61] Ge, M., et al., *A two-dimensional model based on the expansion of physical wake boundary for wind-turbine wakes.* Applied Energy, 2019. **233**: p. 975–984.

[62] Abdelsalam, A.M. and M. El-Shorbagy, *Optimization of wind turbines siting in a wind farm using genetic algorithm based local search.* Renewable Energy, 2018. **123**: p. 748–755.

[63] Veeramachaneni, K., et al., "Optimizing energy output and layout costs for large wind farms using particle swarm optimization," in *Evolutionary Computation (CEC), IEEE Congress,* 2012.

[64] Chowdhury, S., et al., *Unrestricted wind farm layout optimization (UWFLO): Investigating key factors influencing the maximum power generation.* Renewable Energy, 2012. **38**(1): p. 16–30.

[65] Wagner, M., J. Day, and F. Neumann, *A fast and effective local search algorithm for optimizing the placement of wind turbines.* Renewable Energy, 2013. **51**: p. 64–70.

[66] Kusiak, A. and Z. Song, *Design of wind farm layout for maximum wind energy capture.* Renewable energy, 2010. **35**(3): p. 685–694.

[67] Sukumar, S., et al., *Ramp-rate control approach based on dynamic smoothing parameter to mitigate solar PV output fluctuations.* International Journal of Electrical Power & Energy Systems, 2018. **96**: p. 296–305.

[68] Gao, X., et al., *Wind turbine layout optimization using multi-population genetic algorithm and a case study in Hong Kong offshore.* Journal of Wind Engineering and Industrial Aerodynamics, 2015. **139**: p. 89–99.

[69] Ulku, I. and C. Alabas-Uslu, *A new mathematical programming approach to wind farm layout problem under multiple wake effects.* Renewable Energy, 2018.

[70] Eroğlu, Y. and S.U. Seçkiner, *Design of wind farm layout using ant colony algorithm.* Renewable Energy, 2012. **44**: p. 53–62.

[71] Amaral, L. and R. Castro, *Offshore wind farm layout optimization regarding wake effects and electrical losses.* Engineering Applications of Artificial Intelligence, 2017. **60**: p. 26–34.

[72] Khare, V., S. Nema, and P. Baredar, *Solar–wind hybrid renewable energy system: A review.* Renewable and Sustainable Energy Reviews, 2016. **58**: p. 23–33.

[73] Barszcz, T., et al., *Wind speed modelling using Weierstrass function fitted by a genetic algorithm.* Journal of Wind Engineering and Industrial Aerodynamics, 2012. **109**: p. 68–78.

[74] Gonzalez-Longatt, F.M., et al., *Optimal electric network design for a large offshore wind farm based on a modified genetic algorithm approach.* IEEE Systems Journal, 2012. **6**(1): p. 164–172.

[75] Wędzik, A., T. Siewierski, and M. Szypowski, *A new method for simultaneous optimizing of wind farm's network layout and cable cross-sections by MILP optimization.* Applied energy, 2016. **182**: p. 525–538.

[76] Nandigam, M. and S.K. Dhali. "Optimal design of an offshore wind farm layout," in *2008 International Symposium on Power Electronics, Electrical Drives, Automation and Motion*, IEEE. 2008.

[77] Bakhtvar, M., A. Al-Hinai, M.S.E. Moursi, M. Albadi, A. Al-Badi, A.A. Maashri, R.A. Abri, N. Hosseinzadeh, Y. Charaabi, and S. Al-Yahyai, Optimal Scheduling for Dispatchable Renewable Energy Generation. In Proceedings of the 2020 6th IEEE International Energy Conference (ENERGYCon), Tunis, Tunisia, 28 September–1 October 2020, p. 238–243.

[78] Nguyen, C. and H. Lee, *Effective power dispatch capability decision method for a wind-battery hybrid power system.* IET Generation, Transmission & Distribution. 2016, **10**: p. 661–668.

[79] Howlader, A.M., et al., *A review of output power smoothing methods for wind energy conversion systems.* Renewable and Sustainable Energy Reviews, 2013. **26**: p. 135–146.

[80] Hill, C.A., et al., *Battery energy storage for enabling integration of distributed solar power generation.* IEEE Transactions on Smart Grid, 2012. **3**(2): p. 850–857.

[81] Akatsuka, M., et al., "Estimation of battery capacity for suppression of a PV power plant output fluctuation," in *2010 35th IEEE Photovoltaic Specialists Conference*. 2010.

[82] Cheng, F., et al., "Applying battery energy storage to enhance the benefits of photovoltaics," in *2012 IEEE Energytech*. 2012.

[83] Muyeen, S., et al., *Application of energy capacitor system to wind power generation.*, Wind Energy, 2008. **11**(4): p. 335–350.

[84] Kim, T., et al., "A smoothing method for wind power fluctuation using hybrid energy storage," in *2015 IEEE Power and Energy Conference at Illinois (PECI)*. 2015.

[85] Balamurugan, E. and S. Venkatasubramanian, *Analysis of double moving average power smoothing methods for photovoltaic systems.* International Research Journal of Engineering and Technology, 2016. **3**(2): p. 1260–1262.

[86] Sandhu, K.S. and A. Mahesh, *A new approach of sizing battery energy storage system for smoothing the power fluctuations of a PV/wind hybrid system.* International Journal of Energy Research, 2016. **40**(9): p. 1221–1234.

[87] Ellis, A., et al., "PV output smoothing with energy storage," in *38th IEEE Photovoltaic Specialists Conference*. 2012.

[88] Lloyd, C., M. Brown, L. Miller, and A. Von Meier, User requirements and research needs for renewable generation forecasting tools that will meet the needs of the caiso and utilities for 2020. A White Paper Report Prepared by CIEE, 2012.

[89] Holttinen, H., et al., *Impacts of large amounts of wind power on design and operation of power systems, results of IEA collaboration.* Wind Energy, 2011. **14**(2): p. 179–192.

[90] Ye, H., J. Wang, Y. Ge, J. Li, and Z. Li, *Robust Integration of High-Level dispatchable renewables in power system operation.* IEEE Transactions on Sustainable Energy, 2017. **8**: p. 826–835.

[91] Peyghami S., H. Mokhtari, and Blaabjerg, F., "Hierarchical Power Sharing Control in DC Microgrids," in *MICROGRID Advanced Control Methods and Renewable Energy System Integration*, Magdi S. Mahmoud, Ed. Elsevier Ltd, 2017, p. 63–100.

[92] Blakers, A., M. Stocks, B. Lu, C. Cheng, and R. Stocks, *Pathway to 100% Renewable Electricity.* IEEE Journal of Photovoltaics, 2019, **9**: p. 1828–1833.

[93] Chanhom, P., S. Sirisukprasert, and N. Hatti, "A new mitigation strategy for photovoltaic power fluctuation using the hierarchical simple moving average," in *IEEE International Workshop on Inteligent Energy Systems (IWIES)*. 2013.

[94] Alam, M., K. Muttaqi, and D. Sutanto, *a novel approach for ramp-rate control of solar PV using energy storage to mitigate output fluctuations caused by cloud passing.* IEEE Transactions on Energy Conversion, 2014. **29**(2): p. 507–518.

[95] Ernst, B., F. Reyer, and J. Vanzetta, Wind power and photovoltaic prediction tools for balancing and grid operation. In Integration of Wide-Scale Renewable Resources into the Power Delivery System, 2009 CIGRE/IEEE PES Joint Symposium, p. 1–9. IEEE/2009.

[96] Abdullah, M.A., K.M. Muttaqi, D. Sutanto, and A.P. Agalgaonkar, *An effective power dispatch control strategy to improve generation schedulability and supply reliability of a wind farm using a battery energy storage system.* IEEE Transactions on Sustainable Energy, 2015. **6**: p. 1093–1102.

[97] Gholami, M., S.H. Fathi, J. Milimonfared, Z. Chen, and F. Deng, *A new strategy based on hybrid battery–wind power system for wind power dispatching.* IET Generation, Transmission & Distribution, 2018, 12: p. 160–169.

[98] Zhang, F., Z. Hu, K. Meng, L. Ding, and Z.Y. Dong, *Sequence control strategy for hybrid energy storage system for wind smoothing.* IET Generation Transmission & Distribution, 2019, **13**: p. 4482–4490.

[99] Wee, K.W., S.S. Choi, and D.M. Vilathgamuwa, *Design of a least-cost battery-supercapacitor energy storage system for realizing dispatchable wind power.* IEEE Transactions on Sustainable Energy, 2013. **4**: p. 786–796.

[100] Hassaine, L., E. Olias, J. Quintero, and V. Salas, *Overview of power inverter topologies and control structures for grid connected photovoltaic systems.* Renewable and Sustainable Energy Reviews, 2014. **30**: p. 796–807, doi: 10.1016/j.rser.2013.11.005.

[101] Zeb, K. et al., *A comprehensive review on inverter topologies and control strategies for grid connected photovoltaic system.* Renewable and Sustainable Energy Reviews, 2018. **94**(November 2017): p. 1120–1141, doi: 10.1016/j.rser.2018.06.053.

[102] Kumar, A. and S. K. Jain, *A review on the operation of grid integrated doubly fed induction generator.* International Journal of Enhanced Research in Science Technology & Engineering, 2013. **2**(6): p. 25–37.

[103] Eltawil, M.A. and Z. Zhao, *Grid-connected photovoltaic power systems: Technical and potential problems-A review.* Renewable and Sustainable Energy Reviews, 2010. **14**(1): p. 112–129, doi: 10.1016/j.rser.2009.07.015.

[104] T. P. Kumar, N. Subrahmanyam, and M. Sydulu, "Control strategies for a grid-connected hybrid energy system," in *Proceedings of IEEE Conference on TENCONSpring.* 2013. p. 480–484. doi: 10.1109/TENCONSpring.2013.6584491.

[105] Abdullah, M.A., K.M. Muttaqi, D. Sutanto, and A.P. Agalgaonkar, *An effective power dispatch control strategy to improve generation schedulability and supply reliability of a wind farm using a battery energy storage system.* IEEE Transactions on Sustainable Energy, 2015. **6**(3): p. 1093–1102, doi: 10.1109/TSTE.2014.2350980.

[106] Kundur, M.L.P. and N. Balu, *Power System Stability and Control.* New York: McGraw-Hill, 1994.

[107] Bevrani, H., *Robust Power System Frequency Control.* New York: Springer International Publishing, 2009.

[108] Guerrero, M.C.J.M., J.C. Vasquez, J. Matas, and L.G. De Vicuna, *Hierarchical control of droop-controlled AC and DC microgrids—a general approach toward standardization.* IEEE Transactions Industrial Electronics, 2011. **58**(1): p. 158–172.

[109] Guerrero, J.U.J.M., L. Hang, *Control of distributed uninterruptible power supply systems.* IEEE Transactions on Industrial Electronics, 2008. **55**(8): p. 2845–2859.

[110] Hu, J., Y. Shan, Y. Xu, and J. M. Guerrero, *A coordinated control of hybrid AC/DC microgrids with PV-wind-battery under variable generation and load conditions.* International Journal of Electrical Power & Energy Systems, 2019. **104**(April 2018): p. 583–592, doi: 10.1016/j.ijepes.2018.07.037.

[111] Guo, X., Z. Lu, B. Wang, X. Sun, L. Wang, and J. M. Guerrero, *Dynamic phasors-based modeling and stability analysis of droop-controlled inverters for microgrid applications.* IEEE Transactions on Smart Grid, 2014. **5**(6): p. 2980–2987, doi: 10.1109/TSG.2014.2331280.

[112] Magdi S. Mahmoud, "Microgrid Control Problems and Related Issues," in *MICROGRID Advanced Control Methods and Renewable Energy System Integration*, Magdi S. Mahmoud, Ed. Elsevier Ltd, 2017, p. 1–42.

[113] Jamshidi, M.O., *System of Systems Engineering: Principles and Applications.* Boca Raton: CRC Press, 2008.

[114] Zhang, J., et al., *A hybrid method for short-term wind speed forecasting.* Sustainability, 2017. **9**(4): p. 596.

[115] Barthelmie, R.J. and L. Jensen, *Evaluation of wind farm efficiency and wind turbine wakes at the Nysted offshore wind farm.* Wind Energy, 2010. **13**(6): p. 573–586.

[116] MirHassani, S. and A. Yarahmadi, *Wind farm layout optimization under uncertainty.* Renewable Energy, 2017. **107**: p. 288–297.

[117] Jenkins, A., M. Scutariu, and K. Smith, *Offshore wind farm inter-array cable layout.* in *2013 IEEE Grenoble Conference.* 2013. IEEE.

[118] Song, M., et al., *Optimization of wind farm micro-siting for complex terrain using greedy algorithm.* Energy, 2014. **67**: p. 454–459.

6 Renewable Energy Management and Its Dependency on Weather Fluctuation

Abdullah Al Shereiqi, Amer Al-Hinai,
Rashid Al-Abri, Mohammed AlBadi

CONTENTS

6.1 RESEARCH ARCHITECTURE AND DESIGN

A novel optimization strategy is proposed for designing a reliable hybrid plant combining wind, solar, and battery power (HWSPS). The purpose of the strategy is to reduce the power losses typical of wind farms, and at the same time reduce the power fluctuations in the output of wind generation. A genetic algorithm (GA) with a numerical iterative algorithm is used in this study. The proposed strategy is different from existing approaches in that it does not involve a load demand profile. The process of defining the HWSPS capacity is carried out in two main stages. In the first stage, an optimal wind farm is determined using the GA, subject to site dimensions and the spacing between the turbines, taking Jensen's wake-effect model into consideration to eliminate the power losses due to the layout of the wind turbines. At this stage, a sub-case is conducted to co-optimize the wind farm size with cabling connections. The point of common coupling (PCC) is also optimized to be at the border of the selected site. At the second stage, a numerical iterative algorithm is deployed to find the optimal combination of photovoltaic (PV) plant and battery (BESS) sizes in the search space, based on the reference wind power generated by the moving average (MAV), Savitzky-Golay (SG), Gaussian, and locally weighted linear regression (LWLR) techniques. The reliability indices and cost are the basis for obtaining the optimal combination of PV plants and BESS according to a contribution factor with 50 different configurations. A case study is presented to verify the feasibility of the proposed optimal sizing approach.

6.2 WEATHER DATA

This research relies on weather data. The actual data regarding the wind profile, solar irradiance, and temperature are needed to demonstrate the feasibility of the proposed methodology. Instead of using a single input wind profile (i.e., constant speed and constant direction), actual inputs of wind and solar profiles are used here for greater accuracy. The study is based on real wind and solar data recorded in Thumrait, Dhofar Governorate, Oman. The site is called Harweel, and it already has a 50-MW wind farm, as shown in Figure 6.1. It has an average wind speed of around 8 m/s. The weather data are described in the following subsection.

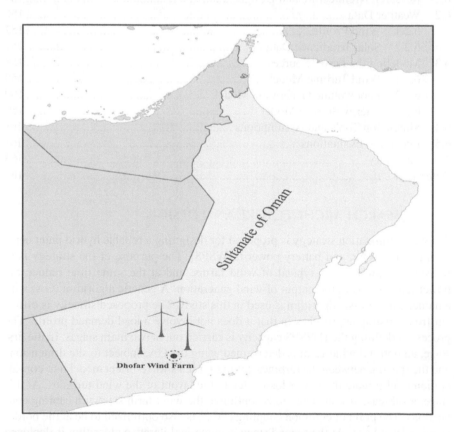

FIGURE 6.1   Location of Dhofar wind farm at the Harweel site.

6.2.1   WIND PROFILE

This study uses wind speed and direction data recorded over 1 year with a resolution of 10-minute windows. The wind profiles were recorded using a wind mast at a height (H_m) of 79 m, but the hub height (H_{wt}) of the wind turbine in this study is 84 m. Thus, to get the wind speed (v_0) with reference to the speed at mast height (v_m), the power law profile expressed in Equation (6.1) is used [1]. The value of wind shear exponent (γ) is commonly taken as (1/7) [2].

$$v_0 = v_m \left(\frac{H_{wt}}{H_m} \right)^{\gamma} \tag{6.1}$$

The velocity deficit depends mainly on wind direction and magnitude. Thus, the wind direction was discretized into 36 segments of 10 degrees each. The probability of the 36 wind speeds occurring is shown in Figure 6.2. This histogram will be used in optimizing the wind farm layout. After determining the farm size, the instant wind power $P_{wind}(t)$ is determined using the actual 10-minute wind profiles, whose wind magnitude and wind direction are shown in Figures 6.3 and 6.4, respectively.

6.2.2   SOLAR IRRADIANCE DATA

To determine the optimal PV plant size, the 10-minute global horizontal irradiance (GHI) and temperature values are used to get the output power of each PV module. Linear interpolation is used to extrapolate for hourly GHI to 10-minute intervals. The GHI profile of the selected site is shown in Figure 6.5. The module's temperature at each time instant also plays a significant role in the solar panel output. The 10-minute temperature profile shown in Figure 6.6 is used.

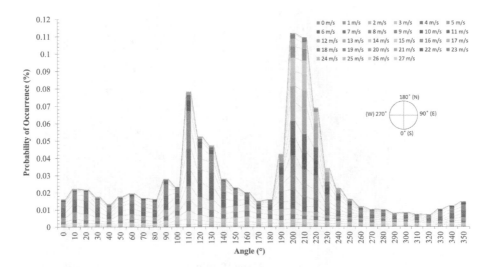

FIGURE 6.2   Histogram of wind speeds and directions for a year.

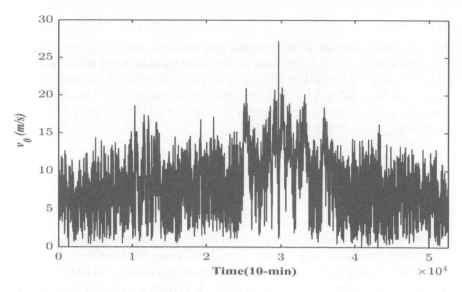

FIGURE 6.3 Wind speed profile for a year.

6.3 MODELING THE POWER SOURCES

The hybrid system uses three power sources: wind turbines, PV modules, and BESS. Modeling the sources is a proactive step in the sizing of the HWSPS. All three models are described in the subsections below.

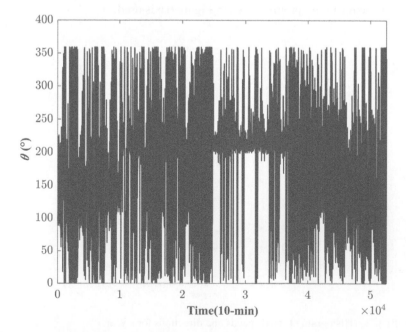

FIGURE 6.4 One-year wind directions.

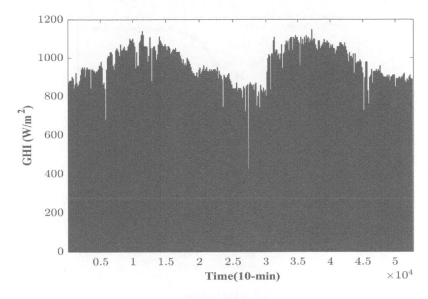

FIGURE 6.5   Global horizontal irradiance for a year.

6.3.1   WIND TURBINE MODEL

The amount of energy harvested from a wind farm is closely linked to the wind turbine installed. In most studies, such as [3], power equations are used to calculate the wind farm's output power. However, it is more precise to linearly interpolate the output power from the power curve supplied by the wind turbine's manufacturer at each input wind speed. In this study, 3-MW wind turbines are used at the wind farm. The output power $P_{wind}(t)$ is the total yield production of all the turbines. Figure 6.7

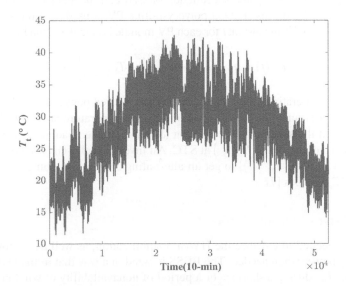

FIGURE 6.6   Temperature at each time instant for a year.

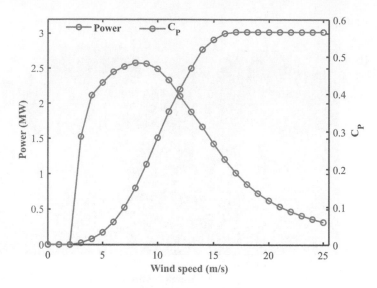

FIGURE 6.7 Power curve of a 3-MW wind turbine.

shows the curves of a wind turbine, including the output power and the power coefficient (C_p) [4]. A constant C_p value does not yield accurate results; hence, variable C_p is used to get a powerful tool for estimating wind farm generation.

6.3.2 Photovoltaic Generation Model

The solar power of a PV plant is a function of GHI and the ambient temperature, as explained in [5, 6]. In this study, a polycrystalline PV is used, and the solar output power $P_{pv}(t)$ at each time instant t for each PV module is calculated via Equation (6.2).

$$P_{pv}(t) = \eta_{pv} P_{rpv} \left[1 + \alpha_{pv} (T_t - T_{STC})\right] \frac{G_t}{G_{STC}} \tag{6.2}$$

where η_{pv} is the derating factor, P_{rpv} is the module's rated power, α_{pv} is the module's temperature coefficient of power, T_t the module's temperature at time t, T_{STC} is the temperature at the standard test condition (STC), G_t is the irradiance at time t, and G_{STC} is the irradiance level $\left(\text{W/m}^2\right)$ at STC. An inverter is chosen based on the maximum power needed $\left(P_{ref}^{max}(t)\right)$ to get an alternating current (AC) output from direct current (DC) sources [7].

6.3.3 Energy Storage Model

A BESS is needed to enhance the HWSPS stabilization, as well as to store energy and supply power when needed. The BESS is sized in a way that mitigates the reference power P_{ref} during a shortage or a period of nonavailability of wind power P_{wind}

or solar power P_{pv}. The BESS state of charge (SOC) is determined according to the wind and solar energy production and reference power, as well as the battery charge and discharge rate, which is designated as $2C$ in this study.

The states of charge and discharge of BESS are calculated as in [8] by defining the BESS energy using Equations (6.3) and (6.4), respectively.

$$E_{B_{ch}}(t) = E_{B_{ch}}(t-1)(1-\delta) + \left[E_{pv}(t) + E_{wind}(t) - \frac{E_{ref}(t)}{\eta_{inv}} \right].\eta_b \qquad (6.3)$$

$$E_{B_{disch}}(t) = E_{B_{disch}}(t-1)(1-\delta) + \left[\frac{E_{ref}(t)}{\eta_{inv}} - E_{pv}(t) - E_{wind}(t) \right].\eta_b \qquad (6.4)$$

where δ is the hourly self-discharge rate, η_{inv} is the efficiency of the inverter, and η_b is the battery charge/discharge efficiency. At each time instant t, the battery's SOC is subject to the following constraints:

$$SOC_{min} \leq SOC(t) \leq SOC_{max} \qquad (6.5)$$

$$SOC_{min} = (1 - DOD)\, C_{battery}\, N_B \qquad (6.6)$$

where the minimum state of charge SOC_{min} is defined using the battery's depth of discharge DOD, and the maximum state of charge SOC_{max} takes the value of the battery's bank capacity $C_{battery}$. A battery with a nominal capacity of $2,000$ Wh and a charge efficiency of 80% is used in this study [9].

6.4 SIMULATION TOOLS AND ASSUMPTIONS

MATLAB® software is used in this study to simulate the proposed approaches. The GA is used to get the optimal wind farm layout. This algorithm is implemented in continuous and discrete search spaces. In addition, a numerical iterative algorithm is used to get the optimal PV plant and BESS sizes. The following assumptions are declared in this book:

- The wind farm size is fixed, with the dimensions of 2×2 km, and the farm is divided into a grid of equal-sized cells.
- The selected site is a flat terrain, and the surface roughness length z_0 is 0.3 m.
- All the wind turbines used in the wind farm are homogenous.
- The turbine nacelle is completely controlled to move the rotor toward the wind direction.
- Spacing between turbines is limited to five rotor diameters.

6.5 ECONOMIC EVALUATIONS

Economic evaluation of the developed system is carried out by considering the system's annual power output for the different dispatching strategies. The average bulk wholesale price of power in the UK electricity market [10] is used to determine the annual incomes for the different strategies. In addition, the accrued benefits from the sale of Renewable Obligation Certificates (ROCs) [10] to power suppliers is calculated and used to determine the payback period for the added benefit of having a battery in the system. This analysis is also compared with the net benefit of avoided cost of reserves for the injected power when considering the conventional approach of off-taking intermittent renewable generation.

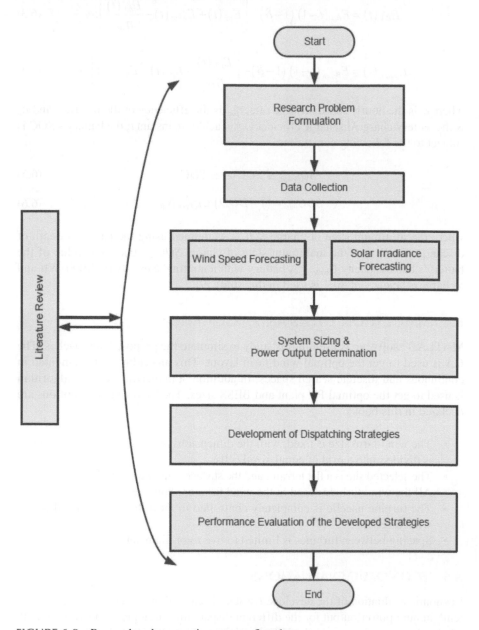

FIGURE 6.8 Research tasks execution process flowchart.

6.6   SUMMARY

This chapter presented the methodology followed in executing the different research tasks. Where required, more clarification on the methods for particular tasks is provided in the next chapters. Figure 6.8 depicts the workflow process that was followed when executing the research tasks. The next chapter delves into development and implementation of the wind speed, solar irradiance, and temperature forecasting models.

REFERENCES

[1] Kikumoto, H., et al., *Observational study of power-law approximation of wind profiles within an urban boundary layer for various wind conditions.* Journal of Wind Engineering and Industrial Aerodynamics, 2017. **164**: p. 13–21.

[2] Valsaraj, P., et al., *Symbolic regression-based improved method for wind speed extrapolation from lower to higher altitudes for wind energy applications.* Applied Energy, 2020. **260**: p. 114270.

[3] Gao, X., H. Yang, and L. Lu, *Study on offshore wind power potential and wind farm optimization in Hong Kong.* Applied Energy, 2014. **130**: p. 519–531.

[4] ENERCON, ENERCON Product Overview. 2015.

[5] Appelbaum, J. and T. Maor, *Dependence of PV module temperature on incident time-dependent solar spectrum.* Applied Sciences, 2020. **10**(3): p. 914.

[6] Dai, Q., J. Liu, and Q. Wei, *Optimal photovoltaic/battery energy storage/electric vehicle charging station design based on multi-agent particle swarm optimization algorithm.* Sustainability, 2019. **11**(7): p. 1973.

[7] Singh, S., M. Singh, and S.C. Kaushik, *Feasibility study of an islanded microgrid in rural area consisting of PV, wind, biomass and battery energy storage system.* Energy Conversion and Management, 2016. **128**: p. 178–190.

[8] Kaabeche, A., M. Belhamel, and R. Ibtiouen, *Sizing optimization of grid-independent hybrid photovoltaic/wind power generation system.* Energy, 2011. **36**(2): p. 1214–1222.

[9] Ma, T. and M.S. Javed, *Integrated sizing of hybrid PV-wind-battery system for remote island considering the saturation of each renewable energy resource.* Energy Conversion and Management, 2019. **182**: p. 178–190.

[10] https://www.ofgem.gov.uk/publications-and-updates/wholesale-energy-markets-2015. Accessed: 2016-02-15.

7 Sizing of the Hybrid Renewable-Based Power Systems

Abdullah Al Shereiqi, Amer Al-Hinai, Rashid Al-Abri, Mohammed AlBadi

CONTENTS

7.1 OPTIMAL SIZING METHODOLOGY

Because the objective is an optimal combination of wind, photovoltaics (PVs), and battery energy storage system (BESS), the study developed a methodology for an optimal Hybrid Wind-Solar Photovoltaic Power Systems (HWSPS). It involves some factors like the wake-effect model to reduce the power losses in the wind farm. In addition, the smoothing technique is used to mitigate the output fluctuations. Hence, the following subsections describe the optimal HWSPS sizing approach.

7.1.1 WIND FARM SIZING

The sizing approach utilizes the actual recorded data for wind speed and solar irradiance with the optimization techniques to size a suitable wind and PV plant capacity.

DOI: 10.1201/9781003307433-7

This approach is applicable on a site that has rich wind resources with abundant solar irradiance to make use of the complementary nature of the wind and PV resources. Thus, sizing of a wind farm is the primary objective in this study, and it is the basis for determining the sizes of PV and BESS. A genetic algorithm (GA) is used to get the optimal wind farm size. The methodology depends on the dimension of the selected site, the location of the turbines, and the wind speeds and directions. Wind power changes, with little variation in wind speed, due to the cubical proportionality between them [1]. Thus, the wake-effect model is deployed to reduce the power losses due to the wind turbines' layout in the farm. When there is wake effect there is a reduction in the wind speed reaching upstream turbines due to the downstream turbines. This effect is induced as a reduction in wind speeds reaching downstream turbines. Jensen's wake-effect model is one of the models that is widely used due to its simplicity and relatively good accuracy [2]. To illustrate the idea of Jensen's model, Figure 7.1 shows three hypothetical wind turbines. T_p is totally inside the wake of T_i, but T_n is partially affected by the wake of T_i. The wind turbine is affected by the wake effect if one of the conditions mentioned in Equation (7.1) is met:

$$\xi_{w,in} = \begin{cases} 1 \ if \ x_{T,i} - R_{w,in} \leq x_{T,n} - R_{0,T,n} \leq x_{T,i} + R_{w,in} \ and \ y_{T,n} > y_{T,i} \\ 1 \ if \ x_{T,i} - R_{w,in} \leq x_{T,n} + R_{0,T,n} \leq x_{T,i} + R_{w,in} \ and \ y_{T,n} > y_{T,i} \\ otherwise \end{cases} \tag{7.1}$$

For partial shadowing, only a percentage of the turbine's affected area to the wind turbine's swept area is taken. Thus, only the affected area C_{shadow} of T_n is considered in calculating the wake effect due to T_i, as given by Equation (7.2):

$$C_{shadow}$$

$$= \frac{R_{0,T,n}^2 \cos^{-1}\left[\dfrac{x_{T,in}^2 + R_{0,T,n}^2 - R_{w,in}^2}{2x_{T,in}R_{0,T,in}}\right] + R_{w,in}^2 \cos^{-1}\left[\dfrac{x_{T,in}^2 - R_{0,T,n}^2 + R_{w,in}^2}{2x_{T,in}R_{w,in}}\right] - C_{ov,lap}}{\pi R_{0,T,n}^2} \tag{7.2}$$

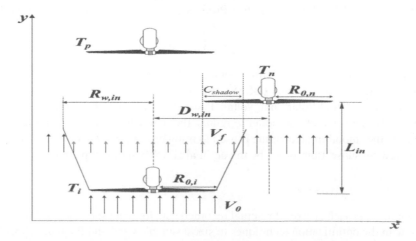

FIGURE 7.1 Wake effect of T_i on T_p and T_n (top view of the wind farm).

where

$C_{ov,lap}$

$$= 0.5 * \sqrt{(-x_{T,in} + R_{0,T,n} + R_{w,in})(x_{T,in} - R_{0,T,n} + R_{w,in})(x_{T,in} + R_{0,T,n} - R_{w,in})(x_{T,in} + R_{0,T,n} + R_{w,in})}$$

(7.3)

Accordingly, the wind speed ($v_{f,n}$) that reaches T_n can be calculated by taking the summation of the wake effects of the other turbines (N) [3]:

$$V_{f,n} = V_o \left[1 - \sqrt{\sum_{\substack{i=1 \\ i \neq n}}^{N} \left[\xi_{w,in} C_{shadow} \frac{\left(1 - \sqrt{1-C_T}\right)}{x_{w,in}} \right]} \right] \quad i,n \in \left[\{1,2,\ldots\ldots,N\}\right] \quad (7.4)$$

Where

$$x_{w,in} = \left[1 + \frac{y_{T,in}}{2 \ln\left(\dfrac{H_T}{z_o}\right)(y_{T,in} + R_{0,T,i})} \right]^2$$

The histograms of wind speeds and directions are the significant inputs to get the optimal output power of the wind farm in a way to reduce the wake effect among the turbines. The main idea in this study is to get the number of turbines (N) that can be allocated in the selected site. It is an optimization problem that produces the coordinates of the turbines' (x,y) cells to get the minimum value of the objective function. The selected site of $2,000 \times 2,000$ m is divided into a grid of 10×10 cells, with a cell size of 200×200 m, in which turbines can be located in the center of the cells. Thus, the total number of variables is 100, representing each cell in the grid. This (WFLO) problem has technical and cost restrictions. The objective is to minimize the cost of energy (COE). The wind farm's power output for each wind scenario is weighted by the scenario's probability $p(V,\theta^w)$, and the expected power output $p_i(x,y,V,\theta^w)$ is incorporated in the optimization objective. This section represents the WFLO model to get the minimal COE as a function of the WT's numbers and locations. The WFLO model is presented mathematically in Equations (7.5) to (7.9):

$$Min(COE) = Min \left[\frac{N\left(\dfrac{2}{3} + \dfrac{1}{3} e^{\left(-0.00174N^2\right)}\right)}{\displaystyle\int_0^{360^o} \int_0^{\infty} \sum_{i=1}^{N} P_i\left(x,y;V,\theta^w\right) p\left(V,\theta^w\right) dV d\theta^w} \right] \quad (7.5)$$

Subject to

$$0 \le x_K \le 1 \quad \forall K \in [i,n], l = 2,000 \tag{7.6}$$

$$0 \le y_K \le w \quad \forall K \in [i,n], w = 2,000 \tag{7.7}$$

$$D_{spec,in} = \sqrt{[x_{T,i} - x_{T,n}]^2 [y_{T,i} - y_{T,n}]^2} \ge (5 \times R_{0,T,n}) \; i,n \in [\{1,2,\ldots,N\}], i \neq n \tag{7.8}$$

$$N > 0 \tag{7.9}$$

The nominator of the objective function in Equation (7.5) represents the economies of scale in wind farms with a constant term and a variable term [4]. The variable term indicates the inverse relationship with the number of wind turbines. In addition, the denominator is used to calculate the average annual energy generated, which is equivalent to calculating the power, because the number of hours per year that a wind farm produces electricity is considered to be constant. Integrating the wind speed from zero to infinity is similar to integrating between cut-in and cut-out speeds. The constraint of the farm's boundaries, as stated in Equations (7.6) and (7.7), ensures placing the wind turbines within the specified area, which is a realistic constraint for most layout problems. In addition, the minimum spacing $D_{spec,in}$ between wind turbines is set to be five times the turbine radius, as represented in Equation (7.8). Moreover, the constraint in Equation (7.9) is to enforce the number of turbines to be greater than zero. The optimization uses 600 populations with 1000 generations to evaluate the fitness value. The optimization process stops when the improvement in the fitness value has been below a threshold for a number of consecutive steps or if the maximum number of generations is reached.

7.1.2 SMOOTHING MODEL

Smoothing of $P_{wind}(t)$ is important in defining the optimal operational policies to limit the ramp rates. Limiting of the generation ramp rate should be addressed carefully to ensure mitigating the output power variability. The ramp rate indicates a power change event in a specific time interval. The ramp rate could be ramp-up (positive) or ramp-down (negative) [5]. It can be supported by power curtailment in the ramp-up rate and limiting the ramp-down rate using the reserve services. Limiting the ramping is crucial for protecting renewable plants and for regulating them easily by other conventional power plants [6].

In this study, the moving average smoothing method (MAV) is used to get a smoothed reference output power $\left(P_{ref}\right)$, which represents the load demand. The value of P_{ref} is the reference for determining the optimal HWSPS plant. According to the literature [7–9], MAV is mainly used to generate a smoother output. The period (k) utilized in MAV to get the P_{ref} value is significant. Thus, the value of k must be defined in a way to reach the desired ramping rate and at the same time to avoid over smoothing of the output power and to keep the ramping within the grid codes regulations. In general, the value of k depends on the wind profile and the grid codes in the selected site. In this study, the ramping is limited to 10% of the wind farm's rated power in a 10-minute time window. The mathematical representation of the k-period MAV is given by Equation (7.10):

$$P_{ref}(t) = \frac{P_{wind}(t) + P_{wind}(t-1) + P_{wind}(t-2) + \ldots + P_{wind}(t-k)}{k} \qquad (7.10)$$

The power fluctuation $\Delta P_{ref}(t)$ is measured at each time instant, taking the difference between the subsequent output powers, as stated in Equation (7.11):

$$\Delta P_{ref}(t) = \left| P_{ref}(t+1) - P_{ref}(t) \right| \qquad (7.11)$$

7.1.3 SIZING OF THE PHOTOVOLTAIC AND BATTERY

The main purpose of utilizing PV and BESS is to get a smoother output power. The optimal size of the wind farm is fixed as determined in Section 7.1.1. To get the optimal combination of PV and BESS in a wind farm, numerical iterative algorithm (NIA) has been employed. First of all, the output power of each PV module was calculated by using Equation (6.2). Then, a contribution factor (S) is defined from 0 to 1, with an increment of 0.01, to get the size of the PV plant. The purpose of using a contribution factor is to determine a range for the search space. Some authors [10] utilized a saturation factor to size the PV and wind plants with an increment of 0.02. However, in this chapter, the determination of PV size is calculated using Equation (7.12) with 100 different solutions. When $S = 0$, there is no need for PV power, and $S = 1$ means that the power generated by the PV plant is equal to $P_{ref}(t)$. Accordingly, there are 100 configurations of PV and BESS in the defined search space:

$$N_{pv}(t) P_{pv}(t) = S P_{ref}(t) \quad S = [0, 0.02, \ldots 1] \tag{7.12}$$

After getting the number of PV modules (N_{pv}), BESS is sized based on the cumulative net energy (CE) according to Equation (7.13). Also, BESS can provide efficient power smoothing without PV but with positive cost implications [11]. Thus, in this study, the BESS is operated as a second source for smoothing purposes to reduce its capacity as well as to maintain the cost of the system:

$$CE_S(t) = CE_S(t-1) + \eta_b E_{gen,S}(t) \tag{7.13}$$

where

$$E_{gen,S}(t) = \left(E_{wind,S}(t) + E_{pv,S}(t) - E_{ref}(t) \right)$$

A positive E_{gen} value means that the BESS is in charge mode, and a negative value means that it is in discharge mode. To avoid oversizing of BESS, the value of CE_S is zero for positive CE_S.

A negative CE_S is considered only for determining the BESS capacity, because the purpose of the BESS is to mitigate any shortage in the generation rather than to store all the excess power. Therefore, the minimum value of $CE_S(t)$ for the whole time-series data is used to determine the capacity $(C_{battery})$ of BESS for each configuration, as stated in Equation (7.14).

$$C_{battery} = \frac{\min\left(CE(t)_{S,year} \right)}{DOD} \tag{7.14}$$

7.2 TECHNO-ECONOMIC MODELING

The primary objective of the integrated approach is to size the HWSPS to get a cost-effective and reliable system. Reliability and system cost analysis play a vital role in determining the HWSPS size. The following subsections describe the techno-economic modeling to evaluate the proposed configurations of the HWSPS.

7.2.1   Cost of Energy

The COE is a key performance indicator for evaluating the proposed HWSPS. Therefore, the COE was calculated for all 50 configurations of wind, PV plant, and BESS to find the minimal COE.

The COE was used to rank the optimal combinations of wind, PV plant, and BESS that satisfy the specified constraints. It is defined as the sum of the capital costs of the generation units in the plant C_{cap}, the operation and maintenance costs C_{OM}, the replacement cost C_{rep}, and the salvage costs C_S, divided by the total energy generated. The COE formula that takes the capital costs, operation and maintenance costs, replacement costs, and residual value of all the components is given in Equation (7.15) [12, 13].

$$COE = \frac{[C_{cap} + C_{O\&M} + C_{rep} - C_S] \; x \; \dfrac{i_r(1+i_r)^{L_{pro}}}{(1+i_r)^{L_{pro}} - 1}}{\displaystyle\sum_{L_s=1}^{L_{pro}} \left[\left(E_{wind} + E_{pv} + E_{disch} \right) x \; \dfrac{1}{(1+i_r)^{L_{pro}}} \right]} \tag{7.15}$$

$$C_{cap} = N\,CC_{wt} + N_{pv}CC_{pv} + N_B CC_B + N_{inv}CC_{inv} \tag{7.16}$$

$$C_{O\&M} = \begin{bmatrix} N \\ N_{PV} \\ N_B \\ N_{inv} \end{bmatrix} \left[\sum_{G_s=1}^{L_{pro}} C_{OM_{wt}} I_r \; \sum_{G_s=1}^{L_{pro}} C_{OM_{pv}} I_r \; \sum_{G_s=1}^{L_{pro}} C_{OM_B} I_r \; \sum_{G_s=1}^{L_{pro}} C_{OM_{inv}} I_r \right] \tag{7.17}$$

where

$$I_r = \frac{1}{(1+i_r)^{G_s}}$$

$$C_{rep} = \sum_{G_s=1}^{Y_{rep}} C_{rep_{unit}} \frac{1}{(1+i_r)^{G_s}} \tag{7.18}$$

$$Y_{rep} = \frac{L_{pro}}{L_{unit}} - 1 \tag{7.19}$$

$$C_S = C_{rep} \; x \; \frac{L_{unit} - \left[L_{pro} - L_{unit} \; \text{int} \left(\dfrac{L_{pro}}{L_{unit}} \right) \right]}{L_{unit}} \tag{7.20}$$

The capital costs CC are the costs pertaining to the supplies and installation work required for each generating unit, as given in Equation (7.16). Yearly operation and maintenance costs $C_{O\&M}$ for each generation unit are used in Equation (7.17) to get the total operation and maintenance costs over the project's lifetime (L_{pro}) while accounting for the discount factor and real interest rate (i_r). Equation (7.18) represents the replacement cost C_{rep}, which is applicable only if the lifetime of the unit L_{unit} is shorter than the lifetime of the project. It is calculated using the replacement cost of the unit $C_{rep_{unit}}$, the discount factor, and the expected number of unit replacements Y_{rep}. The last equation is used to get the salvage cost or residual value, which is a value that the system's components retain at the end of the project's lifetime.

7.2.2   POWER SYSTEM RELIABILITY

Evaluating reliability is vital in the design of power systems. Different reliability indices have been used, such as loss of the power supply probability (LPSP) in [14] and fluctuation rate in [13]. Reliability indices are the antagonists of the COE. Different contributing factors could produce a similar COE; therefore, the reliability indices are used to get higher system reliability within the desired COE.

In this study, the LPSP is used for the techno-economic evaluation and comparison. The LPSP is defined as the probability that the HWSPS cannot fulfill the demand, and its value ranges from 0 to 1. The best LPSP value is 0, which means that the generation fully satisfies the reference power. An LPSP of 1 means that the required reference power is not met [15]. The relative LPSP is defined as

$$LPSP = \frac{\sum\nolimits_{t=1}^{l}\left[[P_{ref}(t) - \left(N_{pv}\ P_{pv}(t) + N\ P_{wind}(t) + P_{disch}(t)\right)\right]}{\sum\nolimits_{t=1}^{l}\left[P_{ref}(t)\right]} \qquad (7.21)$$

where l is the operating time of the HWSPS system, which is equal to 52,560 for a time window of 10 minutes.

7.3   CASE STUDY: EVALUATING THE APPROACH

A case study is presented to demonstrate the usefulness of the proposed approach. Weather data recorded in Thumrait, Dhofar Governorate, Oman, are used, as explained in Chapter 6. The main technical and economic parameters used in this study for the PV module, wind turbines, and BESS are listed in Appendix A. This case study is simulated in MATLAB® following the steps of the proposed approach for sizing the HWSPS, as illustrated in Figure 7.2.

7.3.1   RESULTS AND DISCUSSION

The previous sections describe the approach to sizing the HWSPS. The approach was divided into two main stages. The first stage was to find the optimal WFLO, accounting for the wake-effect model.

The second stage was to find the optimal size of the PV plant and BESS, based on the smoothed wind power output. To show the impact of the wake effect more clearly on wind power generation, Figure 7.3 shows the wind farm's power generation with and without the wake effect using the optimal layout. Applying the wake effect model gives more accurate and realistic results for a wind farm's power generation [16]. The total velocity loss due to the wake effect on the turbines is around 5.36% of the available wind speed. This corresponds to a loss of around 10.48% in the wind

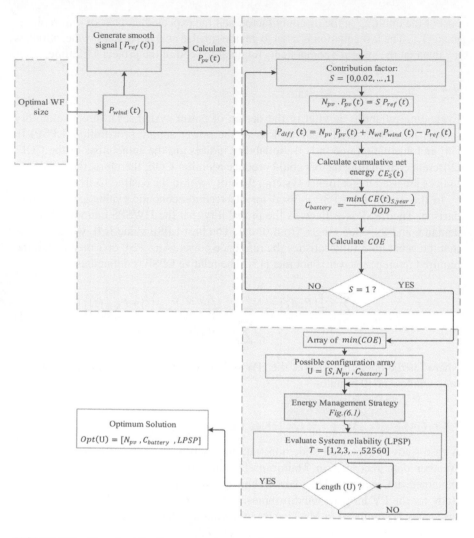

FIGURE 7.2 Flowchart for the proposed approach of HWSPS.

farm's generation power. The difference between the two scenarios is clear, and using the wake-effect model allows planners to project more realistic wind farm generation.

The next step in the proposed approach is to generate a smooth wind output power (P_{ref}) using the locally weighted linear regression (LWLR), MAV, Gaussian, and Savitzky-Golay (SG) to reduce the fluctuations in the wind farm's output power with the aid of a PV plant and BESS. The main purpose of smoothing the wind generation is to reduce the ramping rate, which is clearly shown in Figure 7.4. Ramping rates before and after smoothing are shown. The maximum ramping before smoothing is 30.03 MW. After smoothing, the maximum ramping rates are 2.1 MW (with LWLR), 3.6 MW (with MAV), 2.34 MW (with Gaussian), and 2.98 MW (with SG). That means that the maximum ramping in $P_{refLWLR}$, P_{refMAV}, $P_{refGaussian}$, and P_{refSG} correspond to

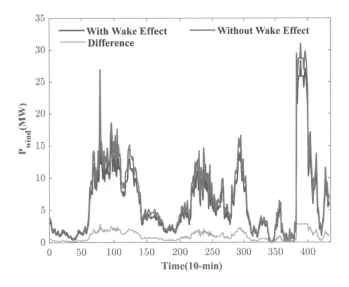

FIGURE 7.3 The wind farm's output power for three days.

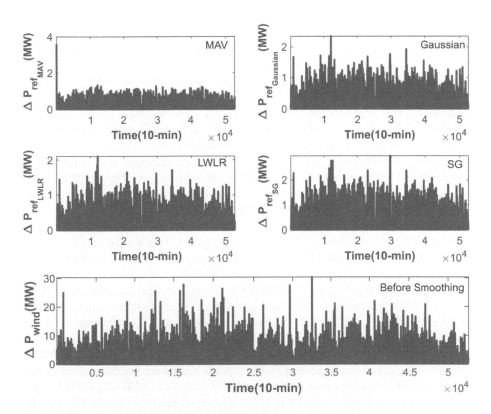

FIGURE 7.4 The wind farm's annual ramping rate.

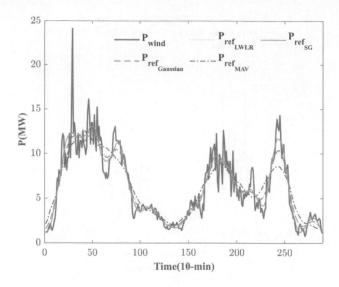

FIGURE 7.5 Sample of actual and smoothed wind power.

5.38%, 9.23%, 6%, and 7.6% of the wind farm's capacity, respectively. These results are sufficient to proceed with this case study.

Each smoothing technique was used with 30 data points, and the generated P_{ref} for each technique is presented in Figure 7.5. It is clear that P_{ref} is smoother than the actual P_{wind}. The purple-bounded areas in Figure 7.6 show how much P_{wind} falls short of P_{ref}. The gap between P_{wind} and P_{refMAV} is greater than that from other smoothing techniques. This is mainly due to the memory effect of the MAV. That means more resources are needed to compensate the generations. Thus, to mitigate this shortage, a PV plant and BESS were added with 50 contribution factors (S).

Running the proposed approach in NIA yielded different configurations of PV plant and BESS with COE and LPSP, as shown in Figure 7.7. The BESS size decreased with each increment of S for every smoothing technique, and then the size was fixed, because the sizing of BESS is designed to avoid oversizing by considering only the minimum annual negative accumulative net energy (CE), using Equation (7.14). The minimum CE values for the whole year are 0.69988 MWh (for LWLR), 0.62003 MWh (for MAV), 0.56051 MWh (for Gaussian), and 0.40655 MWh (for SG), as shown in Figure 7.8.

The BESS capacity began at 18.06 MWh for $S = 0$, and then reached the minimum value of 0.7750 MWh for $S = 0.04$ (with the MAV). In addition, the minimum COE was 0.023 €/kWh, with 10.083 MW for the PV plant sizes. The simulation results also attest to the effect of LWLR, Gaussian, and SG in smoothing the wind power to overcome the negative impact of the memory effect in MAV. Compared with MAV, the optimal sizes of the BESS and PV for LWLR, Gaussian, and SG were attained while $S = 0.01$. Thus, the optimal COE was 0.0244 €/kWh (for LWLR), 0.021 €/kWh (for Gaussian), and 0.0171 €/kWh (for SG). However, the COE for LWLR was the highest, due to the size of the BESS, which reached 0.8749 MWh.

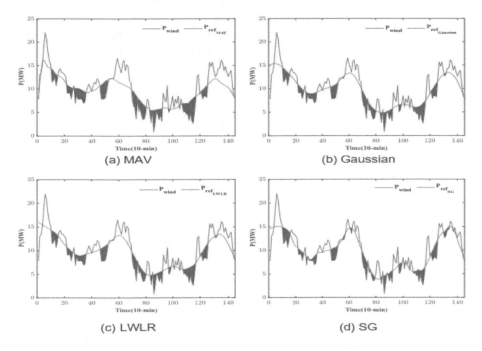

FIGURE 7.6 One-day sample for shortages of P_{wind} to fulfill the P_{ref}; (a) MAV; (b) Gaussian; (c) LWLR; (d) SG.

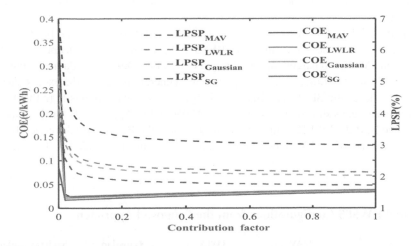

FIGURE 7.7 COE and LPSP for different contribution factors.

On the other hand, the BESS for the SG was 0.1924 MWh lower than the BESS size for the Gaussian. The PV size was 5.01 MW (for LWLR), 5.02 MW (for Gaussian), and 4.99 MW (for SG). These techniques yielded similar PV size values, in contrast with MAV, which yielded a value of 10.08 MW. A summary of the optimum results is shown in Table 7.1.

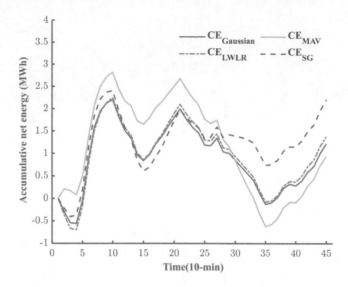

FIGURE 7.8 Minimum accumulative net energy.

As shown in Figure 7.9, the values still fall short of the reference power, but the entire LPSP values for the entire year are 4.05% for MAV, 3.22% for LWLR, 3.03% for Gaussian, and 2.56% for SG. The greatest deficit occurs at nighttime, in the absence of PV generation. The size of the PV plant increases linearly with the increase in a contribution factor S, which leads to higher fluctuations and more damped power.

The power deficit decreases with an increase in a contribution factor, but the COE also increases. At the same time, the BESS SOC fluctuates significantly between SOC_{min} and SOC_{max} (charged and discharged condition). During the daytime, there is sufficient P_{pvMAV} to mitigate the shortage in wind generation. Part of the excess energy is used to charge the BESS. The BESS is charged and discharged mostly in the morning and evening, as shown in Figure 7.10 for MAV and in Appendix B for the remaining methods. If the LWLR, Gaussian, or SG are used, the BESS is used frequently due to the small size of the PV plant. The BESS is used the most with LWLR, followed by

TABLE 7.1

Optimal HWSPS Configurations from the Proposed Approach

	MAV	LWLR	Gaussian	Savitzky-Golay
Wind farm (MW)	39	39	39	39
$C_{battery}$ (MWh)	0.7750	0.8749	0.7006	0.5082
PV (MW)	10.08	5.01	5.02	4.99
COE (€/kWh)	0.023	0.0244	0.0210	0.0171
LPSP (%)	4.05	3.22	3.03	2.56
S	0.04	0.02	0.02	0.02

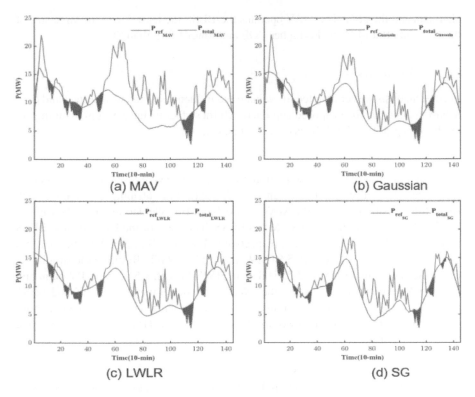

FIGURE 7.9 Sample of total output power and smoothed wind power.

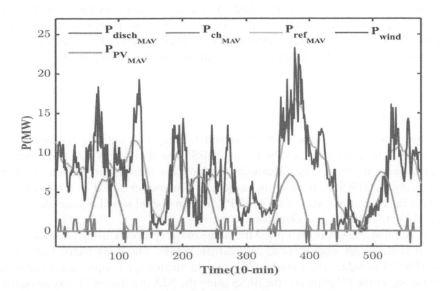

FIGURE 7.10 Sample of four-day HWSPS output power profiles for MAV.

Gaussian and then SG. The MAV operates the BESS the least due to the large size of the PV plant. But MAV yields higher COE and lower LPSP than SG.

7.3.2 SENSITIVITY ANALYSIS

Further analysis was conducted to evaluate the influences of the wake effect and contribution factor on the proposed approach. All these factors have a significant impact on the sizing of the HWSPS, as explained below.

7.3.2.1 Sizing without the Wake Effect

A simulation was carried out to find the optimal HWSPS without considering the wake effect. The simulation ran with the optimal contribution factor for each smoothing technique. With the MAV at $S = 0.04$, this yielded an HWSPS with an 11.37-MW PV plant and a 0.8984-MWh BESS. The sizes of the PV plant and BESS increased by 12.78% and 15.92%, respectively. For the SG at $S = 0.02$, the optimal sizes of the PV and BESS were 5.63 MW and 0.5320 MWh. Running the simulation using the Gaussian with $S = 0.02$ yielded an HWSPS with a 5.66-MW PV plant and a 0.7418-MWh BESS. Finally, for the LWLR technique, the sizes of PV and BESS increased by 11.48% and 6.3% compared with the results with wake. This demonstrates the overestimation that occurs in calculating the HWSPS size if the wake effect is not considered.

7.3.2.2 Effect of Contribution Factor on Sizes of the PV Plant and BESS

The cost of the wind farm alone is lower than the optimum cost of the proposed HWPS. However, the LPSP of an HWSPS is 4.05% with MAV, compared with 8.42% for a stand-alone wind farm, which means that the reliability of the HWSPS is about 4.37% higher. This shows that a wind farm alone is not a reliable system.

Increasing the sizes of the PV plant and the BESS improves the LPSP, but the COE also increases. In addition, the excess power also increases, leading to a reduction in revenue. Thus, the availability of other resources on a wind farm has a positive impact on the electrical system, but only up to a certain level of resource penetration.

7.4 SUMMARY

In this chapter, an approach was developed to design an optimal HWSPS using GA-NIA. The approach sizes the PV and BESS based on the wind farm's output. Unlike previous studies, this approach does not take the load profile into account for sizing the HWSPS. It takes the smoothed wind power signal as a reference to determine the optimal size of the HWSPS. The smoothed signal is generated using the MAV, LWLR, SG, and Gaussian.

In addition, the GA is used in the first stage of the approach, using Jensen's wake-effect model to get more accurate and realistic results. The study focused on sizing the HWSPS to reduce the wind farm's output fluctuations and improve its reliability. The sizing of the PV plant and the BESS using the NIA is a tradeoff between system cost and reliability.

The proposed approach was demonstrated using real global horizontal irradiance (GHI) data from a site in the Sultanate of Oman with a multispeed and multidirectional wind profile. The wake effect on turbines was demonstrated, and the wind farm's power generation with and without the wake-effect model were compared. The optimal HWSPS had a wind to PV ratio of 3.87:1 with MAV and around 7.8:1 with other smoothing techniques. The corresponding BESS capacity represented 1.58% of the HWSPS's rating for MAV, 1.98% for LWLR, 1.16% for SG, and 1.59% for Gaussian.

The results also show that the SG is more suitable for this task than MAV, Gaussian, or LWLR. It was also found that the window size plays a vital role in the smoothing of noisy output, but this smoothness has a positive impact in the cost of system. The memory effect feature of the MAV led to a delay between the smoothed signal and the actual signal. In general, this implies an increase in the smoothing source capacities.

In addition, a sensitivity analysis was conducted to evaluate the influence of the wake effect and the contribution factor on the sizing of the HWSPS. This analysis revealed the importance of the wake effect in avoiding overestimating the HWSPS size. Thus, the proposed approach is effective overall in performing a feasibility study for the sizing of a HWSPS with minimum energy costs.

In the next chapter, the forecasting stage will be explained in detail.

REFERENCES

[1] Shajari, S.; Pour, R.K. Reduction of energy storage system for smoothing hybrid wind-PV power fluctuation. In 2012 11th International Conference on Environment and Electrical Engineering; IEEE: Piscataway, NJ, USA, 2012; pp. 115–117.

[2] Tao, S.; Xu, Q.; Feijóo, A.; Kuenzel, S.; Bokde, N. Integrated wind farm power curve and power curve distribution function considering the wake effect and terrain gradient. Energies 2019, 12, 2482.

[3] Gao, X.; Yang, H.; Lin, L.; Koo, P. Wind turbine layout optimization using multi-population genetic algorithm and a case study in Hong Kong offshore. J. Wind Eng. Ind. Aerodyn. 2015, 139, 89–99.

[4] Mosetti, G.; Poloni, C.; Diviacco, B. Optimization of wind turbine positioning in large windfarms by means of a genetic algorithm. J. Wind Eng. Ind. Aerodyn. 1994, 51, 105–116.

[5] Lee, D.; Kim, J.; Baldick, R. Ramp Rates Control of Wind Power Output Using a Storage System and Gaussian Processes; University of Texas at Austin, Electrical and Computer Engineering: Austin, TX, USA, 2012.

[6] Feng, L.; Zhang, J.; Li, G.; Zhang, B. Cost reduction of a hybrid energy storage system considering correlation between wind and PV power. Prot. Control Mod. Power Syst. 2016, 1, 11.

[7] Sukumar, S.; Mokhlis, H.; Mekhilef, S.; Karimi, M.; Raza, S. Ramp-rate control approach based on dynamic smoothing parameter to mitigate solar PV output fluctuations. Int. J. Electr. Power Energy Syst. 2018, 96, 296–305.

[8] Lam, R.K.; Yeh, H.G. PV ramp limiting controls with adaptive smoothing filter through a battery energy storage system. In Proceedings of the 2014 IEEE Green Energy and Systems Conference (IGESC), Long Beach, CA, USA, 24 November 2014; IEEE: Piscataway, NJ, USA, 2014; pp. 55–60.

[9] Lee, H.J.; Choi, J.Y.; Park, G.S.; Oh, K.S.; Won, D.J. Renewable integration algorithm to compensate PV power using battery energy storage system. In Proceedings of the 2017 6th International Youth Conference on Energy (IYCE), Budapest, Hungary, 21–24 June 2017; IEEE: Piscataway, NJ, USA, **2017**; pp. 1–6.

[10] Ma, T.; Javed, M.S. Integrated sizing of hybrid PV-wind-battery system for remote island considering the saturation of each renewable energy resource. Energy Convers. Manag. **2019**, 182, 178–190.

[11] Howlader, A.M.; Urasaki, N.; Yona, A.; Senjyu, T.; Saber, A.Y. A review of output power smoothing methods for wind energy conversion systems. Renew. Sustain. Energy Rev. **2013**, 26, 135–146.

[12] Shereiqi, A.A., et al., Optimal sizing of a hybrid wind-photovoltaic-battery plant to mitigate output fluctuations in a grid-connected system. Energies. **2020**, 13(11), 3015.

[13] Hatata, A.; Osman G.; M. Aladl. An optimization method for sizing a solar/wind/battery hybrid power system based on the artificial immune system. Sustain. Energy Technol. Assess. **2018**, 27, 83–93.

[14] Fathima, A., et al., Intelligence-based battery management and economic analysis of an optimized dual-vanadium redox battery (VRB) for a Wind-PV HYBRID SYSTEM. Energies. **2018**, 11(10), 2785.

[15] Xu, L., et al., An improved optimal sizing method for wind-solar-battery hybrid power system. IEEE Trans. Sustain. Energy **2013**, 4(3), 774–785.

[16] Tao, S., et al., Optimal micro-siting of wind turbines in an offshore wind farm using Frandsen–Gaussian wake model. IEEE Trans. Power Syst. **2019**, 34(6), 4944–4954.

APPENDIX 7.A: PARAMETERS OF WIND TURBINE, PV MODULE, AND BESS

TABLE 7.A.1

Specifications of the Used Wind Turbine, PV Module, BESS, and Inverter

Wind Turbine		PV	
Rated power	3 MW	Model	Polycrystalline
Hub height	84 m	Maximum power at STC (P_{rpv})	275 W
Rotor diameter	82 m	Temperature coefficient of (P_{rpv})	−0.47%/°C
Capital cost	1784 €/kW	Capital cost	598.62 €/kW
O&M cost	3% capital cost/year	O&M cost	1% capital cost/year
Lifetime	20 years	Lifetime	20 years
BESS		**Inverter**	
Rated power	1000 Ah	Rated power	115 kW
Nominal voltage	2 V	Efficiency (η_{inv})	90%
Capital cost	213 €/kW	Capital cost	117.26 €/kW
Replacement cost	213 €/kW	Replacement cost	117.26 €/kW
O&M cost	9.8 €/kW/year	O&M cost	0.92 €/kW/year
Lifetime	5 years	Lifetime	20 years

APPENDIX 7.B: HWSPS OUTPUT POWER PROFILES

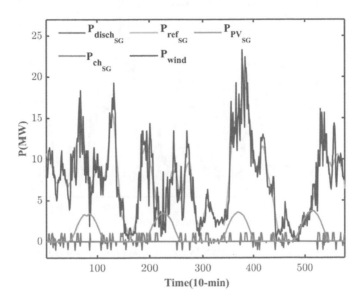

FIGURE 7.B.1   Sample of five-day HWSPS output power profiles using SG.

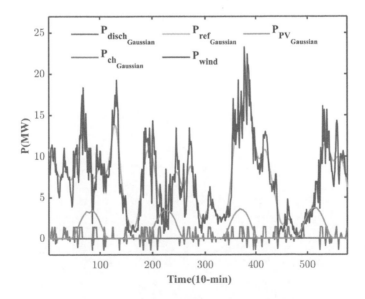

FIGURE 7.B.2   Sample of five-day HWSPS output power profiles using Gaussian.

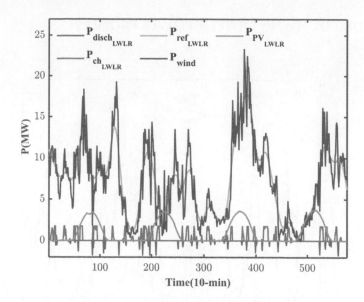

FIGURE 7.B.3 Sample of five-day HWSPS output power profiles using LWLR.

8 Forecasting Tools for Management of Hybrid Renewable Energy Systems

*Saira Al-Zadjali, Ahmed Al Maashri,
Amer Al-Hinai, Sultan Al Yahyai,
Edward Baleke Ssekulima*

CONTENTS

8.1 WIND SPEED FORECASTING

Wind speed forecasting were carried out by implementing a Nonlinear Auto-Regressive with eXogenous input (NARX)-based prediction model to account for its inherent highly stochastic nature.

8.1.1 NARX-BASED ANN MODEL

NARX networks can be designed based on either the parallel architecture or the series-parallel architecture. In the parallel architecture, the predicted output of the previous step is fed back to be used as the input to the feed-forward neural network (NN) for the next forecast. This chapter adopts the series-parallel architecture in which the true value of the previous step is used as the feed-forward input since they can be accessed during the network training. This results in more accurate predictions because the input is more accurate. The resulting NARX network is thus a purely feed-forward configuration. On completion of network training there is a switch to the parallel structure to perform the multistep ahead forecasts.

DOI: 10.1201/9781003307433-8

183

A NARX model is defined by Equation (8.1) [1].

$$\hat{y}_i = f\left(\hat{y}_{i-1}, \ldots, \hat{y}_{i-n_y}, x_{i-1}, \ldots, x_{i-n_x}\right) \tag{8.1}$$

where x_i and y_i represent the inputs (exogenous variables) and outputs of the model, respectively, at time instant, i; f is a nonlinear mapping function that is approximated using a feed-forward NN such as a multilayer perceptron (MLP). The MLP approximation of f comprises a set of nodes organized into H input and J output layers (nodes). An example of a NARX network topology consisting of a feedback element and one hidden layer is shown in Figure 8.1 [2]. The nodes of the input layer, m, correspond to a set of two tapped delay lines. One line contains n_x taps of the exogenous inputs, whereas the other consists of n_y taps of the network outputs. The input nodes are updated using Equation (8.2).

$$m_{i+1}^k = \begin{cases} x_i : k = n_x \\ \hat{y}_i : k = n_x + n_y \\ m_{i+1}^k : 1 \le k < n_x \text{ or } n_x < k < \left(n_x + n_y\right) \end{cases} \tag{8.2}$$

Such that the value of the taps at time instant i, are given by:

$$m_i = \left[x_{i-n_x} \ldots x_{i-1} \, \hat{y}_{i-n_y} \ldots \hat{y}_{i-1} \right] \tag{8.3}$$

The activation function of the hidden layers is defined by Equation (8.4).

$$z_i^p = \sigma\left(\sum_{k=1}^{H} v^{pk} m_i^k + c^p \right) : p = 1, \ldots, J \tag{8.4}$$

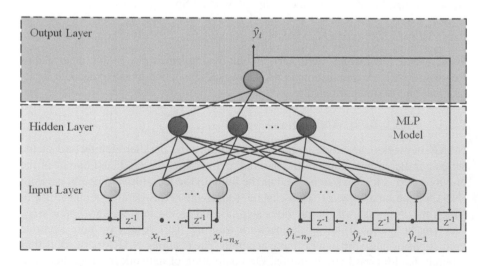

FIGURE 8.1 NARX network setup.

where σ is a nonlinear bounded one-side saturated sigmoid function defined by Equation (8.5).

$$\sigma(x) = \begin{cases} 0 & x \le s \\ \dfrac{1}{1+e^{-x}} & otherwise \end{cases} \tag{8.5}$$

where $s \in \mathcal{R}$, and $v^{p,k} \in V_p$ are input weights and $c^p \in C_p w$ are the hidden layer biases, expressed by Equation (8.6).

$$V_p = \begin{bmatrix} v_P^{1,1} & \cdots & v_P^{1,H} \\ \vdots & \ddots & \vdots \\ v_P^{J,1} & \cdots & v_P^{J,H} \end{bmatrix}; C_p = \begin{bmatrix} c_P^1 \\ \vdots \\ c_P^J \end{bmatrix} \tag{8.6}$$

The output layer contains a single linear node denoted by Equation (8.7)

$$\widehat{y_i} = \sum_{p=1}^{J} v_o^P z_i^p + \Phi^o \tag{8.7}$$

where Φ^o is the output bias term and $v^p \in V_o$ are the output weights given by Equation (8.8)

$$V_p = \begin{bmatrix} v_o^1 \\ \vdots \\ v_o^J \end{bmatrix} \tag{8.8}$$

In this study, the NARX network was set up using the data acquired by measuring 10-minute interval wind speeds over 2 years. The data were divided into training, validation, and testing datasets using a ratio of 8:1:1. The optimal number of hidden layers was determined to be 10 with the feedback delay set to 2. Since the generation must be scheduled 1 hour ahead of the trading period, a three-step ahead prediction of 10-minute interval wind speeds was carried out.

8.1.2 CONSTRUCTING PREDICTION INTERVALS

For the developed proportional-integrals (PIs) to be reliable, the values of PI coverage probability (PICP) should be close to its prescribed nominal coverage rate. The bounds (upper and lower) are calculated using Equations (8.9) and (8.10), where z is the critical value of the standard Gaussian distribution that corresponds to the $100(1-\gamma)\%$ confidence level. This is known as the prediction interval nominal confidence (PINC). A 95% PINC was chosen for this study. The pairs bootstrapping technique was used to develop the bounds. It is basically a statistical inference method that uses data resampling to minimize the standard deviation to keep the bounds as close as possible [3].

$$U_i^t = \hat{y}_i + \frac{z(1-\gamma)}{2}\sqrt{\sigma_{t_i}^2} \tag{8.9}$$

$$L_i^t = \hat{y}_i - \frac{z(1-\gamma)}{2}\sqrt{\sigma_{t_i}^2} \tag{8.10}$$

In instrumentation and measurement of processes, the measured variable or target is represented as:

$$t_i = y_i + \varepsilon_i \tag{8.11}$$

where y_i is the true regression mean and ε is an additive combination of random noise and measurement error with zero expectation. The artificial neural network (ANN) output, $\widehat{y}_i$, is the estimate of the true regression mean;

$$\hat{y}_i = E(t_i \mid x_i) \tag{8.12}$$

As a result, the prediction error is easily expressed as

$$t_i - \hat{y}_i = \left[y_i - \hat{y}_i\right] + \varepsilon_i \tag{8.13}$$

The total variance of the developed prediction model as given by Equation (8.14) is obtained by assuming statistical independence between both terms on the right-hand side of Equation (8.13).

$$\sigma_{t_i}^2 = \sigma_{\hat{y}_i}^2 + \sigma_{\varepsilon_i}^2 \tag{8.14}$$

where $\sigma_{\hat{y}_i}^2$ represents the variance of the model misspecification and $\sigma_{\varepsilon_i}^2$ represents the measure of the data noise variance. The PIs are then developed based on the calculation of these variances. The pairs bootstrapping technique was used to develop the bounds in this chapter. It is basically a statistical inference method that uses data resampling to minimize the standard deviation so as to keep the bounds as close as possible. In other words, an ensemble of NN models would produce a less biased estimate of the true regression of the target output [3]. Starting with the original dataset $D = \left\{(x_i, t_i)\right\}_{i=1}^{N}$, D_M, training datasets are resampled with replacement. In this chapter, 200 D_M NN model replicates are constructed to determine the model misspecification variance, $\sigma_{\hat{y}_i}^2$. The true regression mean is estimated by averaging the point predictions, $\hat{y}_i^n$, of D_M bootstrap models.

$$\hat{y}_i = \frac{1}{D_M} \sum_{k=1}^{D_M} \hat{y}_i^n \tag{8.15}$$

Thereafter, Equation (8.16) is used to estimate $\sigma_{\hat{y}_i}^2$.

$$\sigma_{\hat{y}_i}^2 = \frac{1}{D_M - 1} \sum_{k=1}^{D_M} (\hat{y}_i^n - \hat{y}_i)^2 \tag{8.16}$$

The challenge then lies in estimating the data noise variance, $\sigma_{\varepsilon_i}^2$, because there is only one observation of wind speed available in addition to the heteroscedasticity of the data. Then, $\sigma_{\varepsilon_i}^2$ is expressed by Equation (8.17), thus, the variance squared residuals, r_i, are calculated using Equation (8.18).

$$\sigma_{\varepsilon_i}^2 = E\left[\left(t_i - \hat{y}_i\right)^2 x_i\right] - \sigma_{\hat{y}_i}^2 \tag{8.17}$$

$$r_i = \max\left(\left(t_i - \hat{y}_i\right)^2 - \sigma_{\hat{y}_i}^2, 0\right) \tag{8.18}$$

While holding the input x_i, a new dataset $D_\varepsilon = \left\{(x_i, t_i)\right\}_{i=1}^N$, D_ε is obtained using the calculated values of r_i. Another NARX model, NN_ε is then indirectly trained to estimate the unknown values of $\sigma_{\varepsilon_i}^2$. The NN_ε model misspecification variance, $\sigma_{m_i}^2$ is obtained in a manner similar to $\sigma_{\hat{y}_i}^2$ by utilizing the bootstrap technique. D_ε is resampled with replacement to obtain D_F training datasets and subsequently D_F NN_ε models are developed. Let $\hat{m}_i^n$ be the output of the nth NN_ε, then the estimated noise variance of NN_ε is calculated using Equations (8.19) and (8.20). Thus the total noise variance is given by Equation (8.21).

$$\hat{\sigma}_{\varepsilon_i}^2 = \frac{1}{D_F} \sum_{k=1}^{D_F} \hat{m}_i^n \tag{8.19}$$

$$\sigma_{m_i}^2 = \frac{1}{D_F - 1} \sum_{k=1}^{D_F} (\hat{m}_i^n - \hat{\sigma}_{\varepsilon_i}^2)^2 \tag{8.20}$$

$$\sigma_{\varepsilon_i}^2 = \hat{\sigma}_{\varepsilon_i}^2 + \sigma_{m_i}^2 \tag{8.21}$$

Figure. 8.2 illustrates the variation of the wind speed for a sample of the dataset.

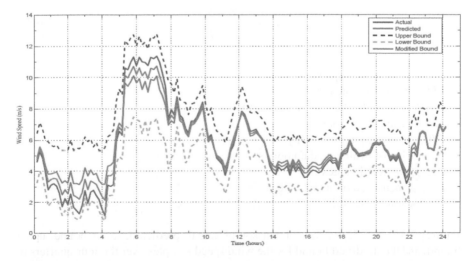

FIGURE 8.2   Wind speed variation for a sample from the first quarter.

Using the model misspecification and data noise variance, the overall variance of the PIs is calculated using Equation (8.14) and the bounds are obtained by Equations (8.9) and (8.10).

8.1.3 MODEL PREDICTION ACCURACY

The NARX model's performance is assessed using Equations (8.22) to (8.27)

$$RMSE = \sqrt{\frac{1}{N}\sum_{k=1}^{N}\left(I_{pred,k} - I_{actual,k}\right)^2} \tag{8.22}$$

$$MAE = \frac{\sum_{k=1}^{N}\left|I_{actual,k} - I_{pred,k}\right|}{N} \tag{8.23}$$

$$MAPE = \frac{100}{N}\times\sum_{k=1}^{N}\left|\frac{I_{actual,k} - I_{pred,k}}{I_{actual,k}}\right| \tag{8.24}$$

$$MBE = \frac{\sum_{k=1}^{N}\left(I_{pred,k} - I_{actual,k}\right)}{N} \tag{8.25}$$

and

$$PICP = \frac{1}{N}\sum_{i=1}^{N}a_i \tag{8.26}$$

where

$$a_i = \begin{cases} 1\ t_i \in I_i^t \\ 1\ t_i \notin I_i^t \end{cases} \tag{8.27}$$

where
   $RMSE$: root mean-square error
   MAE: mean absolute error
   $MAPE$: mean absolute percentage error
   MBE: mean bias error
   $PICP$: PI coverage probability
   N: number of evaluated data points
   $I_{pred,k}$: predicted values
   $I_{actual,k}$: measured or actual values
   I_i^t: stochastic interval

Figures 8.3 to 8.5 depict the comparison of actual, predicted, lower bound, upper bound, and the modified bound for the wind speed samples over the four quarters of the year. Table 8.1 depicts the model performance using the predictions and modified

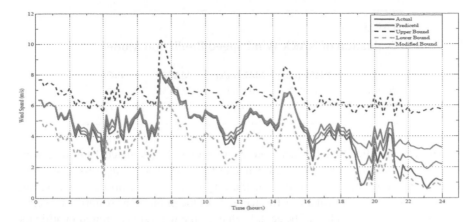

FIGURE 8.3   Wind speed variation for a sample from the second quarter.

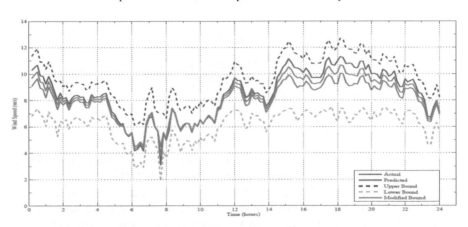

FIGURE 8.4   Wind speed variation for a sample from the third quarter.

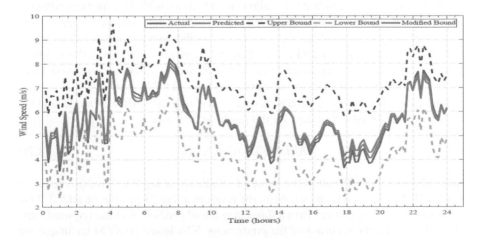

FIGURE 8.5   Wind speed variation for a sample from the fourth quarter.

TABLE 8.1

Evaluation of NARX Wind Speed Prediction Model

	RMSE (m/s)	MAE (m/s)	MAPE	PICP
Predicted	1.4913	0.9471	22.94%	100%
Modified	0.7456	0.4736	11.47%	99.52%

values. The modified bound is achieved by averaging the predicted 10-minute wind speed values at time t and the actual value at $t - 52560$ and taking this as the final prediction that results in a modified bound for dispatching purposes. It is observed that the modified bound enhances the accuracy of the predictions resulting in a close match with the actual values. The actual values of the wind speeds remain within the constructed PI for almost all the intervals.

8.2 SOLAR IRRADIANCE AND TEMPERATURE FORECASTING

Solar irradiance and temperature forecasting was implemented by developing both Seasonal AutoRegressive Integrated Moving Average (SARIMA) and hybrid SARIMA-support vector machine (SVM) models.

8.2.1 SARIMA-SVM-BASED PREDICTION

A SARIMA model incorporates both nonseasonal and seasonal factors in a multiplicative model. The model's shorthand notation is of the form $ARIMA(p,d,q) \times (P,D,Q)_S$ with p = nonseasonal AutoRegressive (AR) order, d = nonseasonal differencing, q = nonseasonal Moving Average (MA) order, P = seasonal AR order, D = seasonal differencing, Q = seasonal MA order, and S = time span of repeating seasonal pattern [4]. Because time series forecasting works best with non-zero data values. The solar irradiance data were segmented into night and day times with daytime being 0800 to 1800 hours. Forecasting was thus carried out for the day data only. After performing a spectral analysis of the data, the appropriate model parameters were determined using the well-known Box-Jenkins method [5]. Equation (8.28) gives the resulting model order. The autocorrelation function (ACF) and partial autocorrelation function (PACF) of the model are shown in Figures 8.6 and 8.7.

Thereafter, two-thirds of the data is used in building the SARIMA model and the remaining third used as test data for the resulting forecasts.

$$ARIMA(3,0,0) \times (3,1,1)_{11} \tag{8.28}$$

Solving Equation (8.28) gives the value of irradiance at lag 0 to have predictors at lags 3, 11, 14, 33, 36, 44, and 47. Also, the residues at lag 33 need to be added. This basically implies we can perform a three-step ahead prediction of the required variable. To enhance the accuracy of the predictions, NNs based on SVM technique are utilized.

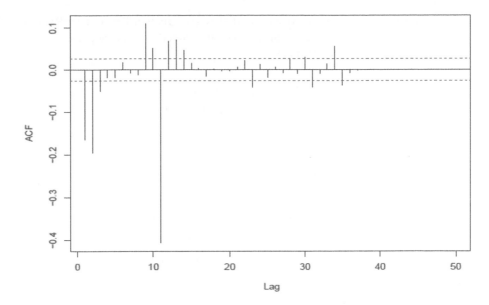

FIGURE 8.6   Solar irradiance ACF.

First, the input data frame is lagged according to Equation (8.29), where data are the hourly solar irradiance values. The SVM model is thus set up using the lagged values. SVM simply refers to a supervised learning algorithm used to implement data classification and analysis. It is basically a discriminative classifier [6]. The data are divided into training, testing, and validation data in the ratio of 7:2:1. The

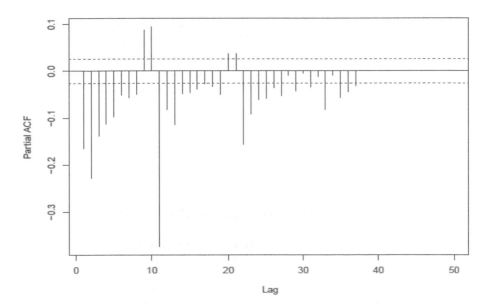

FIGURE 8.7   Solar irradiance PACF.

training is carried out using SVM with Radial Basis Function Kernel (svmRadial) method [7]. Once the model has been successfully trained, a three-step ahead forecast of solar irradiance is carried out.

$$
\left\{
\begin{array}{l}
L < -length(data.\,in) \\
data.\,lag\ 0 < -(data.\,in\,[48\colon (L-0)]) \\
data.\,lag\ 3 < -(data.\,in\,[45\colon (L-3)]) \\
data.\,lag\ 11 < -(data.\,in\,[37\colon (L-11)]) \\
data.\,lag\ 14 < -(data.\,in\,[34\colon (L-14)]) \\
data.\,lag\ 33 < -(data.\,in\,[15\colon (L-33)]) \\
data.\,lag\ 36 < -(data.\,in\,[12\colon (L-36)]) \\
data.\,lag\ 44 < -(data.\,in\,[4\colon (L-44)]) \\
data.\,lag\ 47 < -(data.\,in\,[1\colon (L-47)])
\end{array}
\right.
\tag{8.29}
$$

8.2.2 MODEL PERFORMANCE EVALUATION

As implemented for wind speed prediction, the bootstrapping technique is used to construct the PIs.

Table 8.2 depicts the accuracy parameters achieved when using SARIMA and hybrid SARIMA/SVM models. Figures 8.8 to 8.11 depict the comparison of actual, predicted, lower bound, upper bound, and the modified bound for solar irradiance samples from different quarters of the year for a consecutive period of 4 days between 0800 and 1800 hours.

A similar approach is used to predict the temperature. The temperature model is shown in Equation (8.30). The ACF and PACF of the model are depicted in Figures 8.12 and 8.13. A three-step ahead prediction of temperature is thus carried out following the same procedure described for solar irradiance. However, the training method used is the pcaNET as it achieved convergence faster than SVM and temperature is not a highly stochastic variable.

$$
ARIMA(1,0,2) \times (0,1,0)_{365}
\tag{8.30}
$$

TABLE 8.2

Evaluation of Solar Irradiance Prediction Model

Parameter	SARIMA	SARIMA-SVM
RMSE (W/m^2)	194.8019	54.6521
MAE (W/m^2)	133.7872	34.5973
MAPE	0.55%	0.021%
PICP	100%	100%

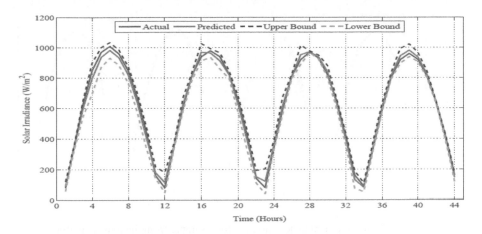

FIGURE 8.8   Daily solar irradiance variation for a sample from the first quarter.

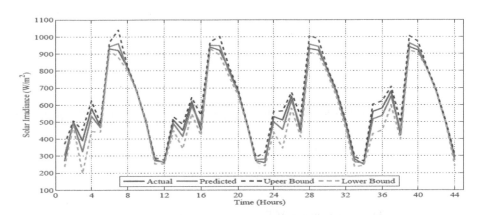

FIGURE 8.9   Daily solar irradiance variation for a sample from the second quarter.

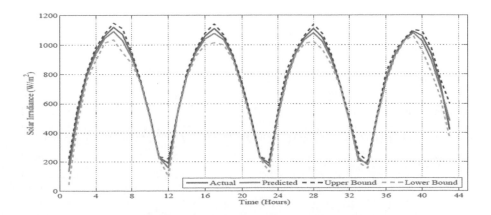

FIGURE 8.10   Daily solar irradiance variation for a sample from the third quarter.

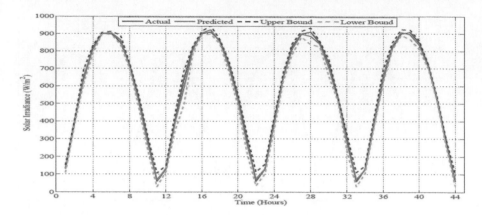

FIGURE 8.11 Daily solar irradiance variation for a sample from the fourth quarter.

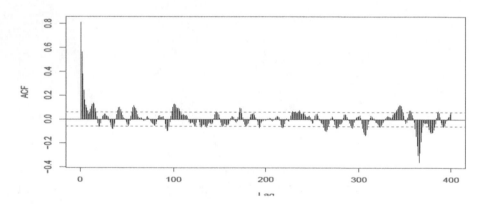

FIGURE 8.12 Temperature model ACF.

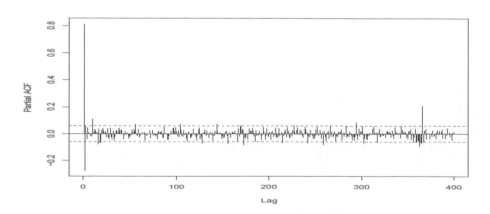

FIGURE 8.13 Temperature model PACF.

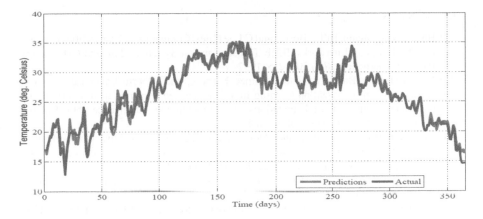

FIGURE 8.14   Daily temperature variation over the year.

Figure 8.14 depicts the predicted and actual daily temperature variation over the year, whereas Table 8.3 shows the evaluation of the developed model. PICP is not evaluated here due to the obvious high level of accuracy. It is imperative to note that the 10-minute interval temperature measurements that were acquired with wind speed data were used in the calculation of power output from the photovoltaic (PV) panels [8].

$$\Delta P_i^{HWSPS} = P_i^{HWSPS,disp} - P_i^{HWSPS} \tag{8.31}$$

8.3   SUMMARY

This chapter presented the modeling of the wind and solar energy resources. The prediction model for wind speed was developed based on NARX NNs utilizing temperature as the exogenous input, whereas that for solar irradiance was developed via implementation of a hybrid SARIMA/SVM model. Subsequently, prediction intervals for wind speeds and solar irradiance were constructed using the bootstrapping technique. The next chapter presents the HWSPS dispatching strategy.

TABLE 8.3
Evaluation of the Temperature Prediction Model

Parameter	SARIMA	SARIMA-SVM
RMSE (°Celsius)	194.8019	54.6521
MAE (°Celsius)	1.8802	0.6544
MAPE	0.02%	0.007%
PICP	100%	100%

REFERENCES

[1] Hava T Siegelmann, Bill G Horne, and C Lee Giles. Computational capabilities of recurrent NARX neural networks. IEEE Transactions on Systems, Man, and Cybernetics, Part B: Cybernetics, 27(2):208–215, 1997.

[2] Kostas Hatalis, Basel Alnajjab, Shalinee Kishore, and Alberto Lamadrid. Adaptive particle swarm optimization learning in a time delayed recurrent neural network for multi-step prediction. In Foundations of Computational Intelligence (FOCI), 2014 IEEE Symposium on, pages 84–91. IEEE, 2014.

[3] Abbas Khosravi, Saeid Nahavandi, Doug Creighton, and Amir F Atiya. Comprehensive review of neural network-based prediction intervals and new advances. IEEE Transactions on Neural Networks, 22(9):1341–1356, 2011.

[4] William Wu-Shyong Wei. Time series analysis, univariate and multivariate methods. Pearson Education, 2006.

[5] George EP Box, Gwilym M Jenkins, Gregory C Reinsel, and Greta M Ljung. Time series analysis: forecasting and control. John Wiley & Sons, 2015.

[6] Jie Shi, Wei-Jen Lee, Yongqian Liu, Yongping Yang, and Peng Wang. Forecasting power output of photovoltaic systems based on weather classification and support vector machines. IEEE Transactions on Industry Applications, 48(3):1064–1069, 2012.

[7] https://r-forge.r-project.org/. Accessed: 2015-06-10.

[8] Jan Kleissl. Solar energy forecasting and resource assessment. Academic Press, 2013.

APPENDIX 8.A: WIND TURBINE AND SOLAR PV MODULE SPECIFICATIONS

The Vestas V110-2.0 MWTM IEC IIIA wind turbine specifications were used in the design. The corresponding turbine power curve is depicted in Figure 8.A.1, whereas Table 8.A.1 details the turbine technical specifications. The PV module technical specifications are presented in Table 8.A.2.

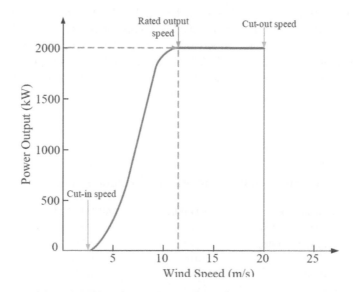

FIGURE 8.A.1 Wind turbine curves.

TABLE 8.A.1
Turbine Technical Specifications

Operational Data

Rated power	2,000 kW
Cut-in wind speed	3 m/s
Cut-out wind speed	20 m/s
Re cut-in wind speed	18 m/s
Wind class	IEC IIIA
Operating temperature range standard turbine	−20C to 40C
Operating temperature range low standard turbine	−30C to 40C

Sound Power

Maximum	107.6 dB

ROTOR

Rotor diameter	110 m
Swept area	9,503 m²
Air brake	Full blade feathering with 3 pitch cylinders

Gear Box

Type	Two planetary stages and one helical stage

Tower

Type	Tubular steel tower
Hub height	80 m (IEC IIIA), 95 m (IEC IIIA,IIIB), 110 m (IEC IIIB), 120 m (IEC IIIB), and 125 m (IEC IIIB)

Nacelle Dimensions

Height for transport	4 m
Height installed (including Cooler top)	5.4 m
Length	10.4 m
Width	3.5 m

Hub Dimensions

Max transport height	3.4 m
Max transport width	4 m
Max transport length	4.2 m

Blade Dimensions

Length	54 m
Maximum chord	3.9 m
Maximum weight per unit for transportation	70 t

TABLE 8.A.2
SANYO HIP-215NHE5 Technical Specifications

Maximum power (P_{max})	[W]	215
Maximum power voltage (V_{pm})	[V]	42.0
Maximum power current (I_{pm})	[A]	5.13
Open circuit voltage (V_{oc})	[V]	51.6
Short circuit current (I_{sc})	[A]	5.61
Warranted minimum power (P_{min})	[W]	204.3
Output power tolerance (P_{max})	[%]	+10/–5
Maximum system voltage	$[V_{dc}]$	1000
Temperature coefficient of (P_{max})	[%/°C]	–0.3
Temperature coefficient of (V_{oc})	[V/°C]	–0.129
Temperature coefficient of (I_{sc})	[mA/°C]	1.68

Note 1: Standard Test Conditions: Air mass = 1.5; irradiance = 100 W/m²; cell temperature = 25°C.
Note 2: The values in Table 8.A.2 are nominal.

9 Dispatching of the Hybrid Renewable Energy Systems

Mostafa Bakhtvar, Amer Al-Hinai

CONTENTS

9.1 DISPATCHABLE HYBRID POWER PLANT

9.1.1 FRAMEWORK

As already introduced in Section 5.1, the management system for dispatchable hybrid renewable power plant with battery energy storage system (DHRB) consists of four key components as shown in Figure 9.1.

- **Forecast unit:** This carries out the forecasting of parameters that affect solar and wind generation, e.g., humidity, temperature, wind speed and direction, and solar irradiance, for a given time interval ahead. Depending on the forecast horizon, the forecasting unit must employ suitable techniques that are sufficiently accurate and fast. Machine learning models, namely nonlinear auto-regressive with exogenous input and support vector machines, among others, exhibit good performance for near-time forecasting [1, 2]. Therefore, they are candidates for employment in the forecasting unit of the DHRB.
- **Aggregation unit:** This uses the forecast data to estimate the maximum power output of the wind power plant and solar photovoltaic (PV) system (P') based on the layout of the power plant, degradation, and other

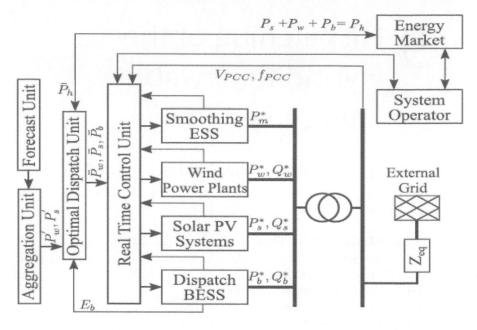

FIGURE 9.1 Framework for DHRB.

characteristics of the equipment, at every given time step. It is emphasized that the output of the forecast and aggregation units should be the forecast power corresponding to the lower confidence limit at a sufficiently high confidence level to enable reliable dispatchable power from DHRB.

- **Optimal dispatch unit:** This finds the optimum set points for the wind power plant, solar PV system, and battery energy storage system (BESS) (P) based on the estimated available power from a wind power plant, solar PV system, and the state of the charge of the BESS such that the equipment (and grid code) constraints are respected; these shall be used to bid in the electricity market. The market decisions $(\bar{P})$ are forwarded to the real-time control unit.
- **Real-time control unit:** Controls the wind power plant, solar PV system, BESS, and smoothing energy storage system (ESS) such that the set points provided by the optimal dispatch unit (based on market decisions) and instructions by the system operator are met (P^*, Q^*).

This chapter focuses on the optimal dispatch unit of the DHRB framework.

9.1.2 Dispatch Time Horizons

Electricity system operators carry out economic dispatch studies ahead of the dispatch time interval (settlement period, d). To ensure optimal operation of the power system, the system operator considers various factors including available generation capacity, cost of generation, demand, and the power system constraints. Therefore, system operators require power generation companies (GENCOs) to submit the volume of the electricity they can provide and the associated cost n_D^- settlement periods ahead of the dispatch (e.g., submission deadline in the UK electricity market is two settlement periods ahead of the dispatch [3]). Moreover, to meet technical constraints and grid code requirements, GENCOs may consider n_D^+ periods after the target settlement period in their resource assessment.

Thus, at every moment, four time horizons ahead may be defined from the GENCO's (in this chapter it is the DHRB operator) perspective for participation in the electricity market (as shown in Figure 9.2 for $n_D^+ = n_D^- = 2$),

- Present operation $\left(d^* - n_D^- - 1\right)$ corresponds to the real time operation of the DHRB, based on the committed power, and is dealt with the real-time control unit (Figure 9.1).
- Past announced settlement periods $\left(d^* - n_D^-\ to\ d^* - 1\right)$ happen when the DHRB operator has already bid in the electricity market and should realize the committed power despite the changes that might have occurred in the available renewable generation forecast to avoid power shortage penalty.
- Target settlement period (d^*) is the nearest (in time) settlement period for which the DHRB operator shall bid.
- Future expected settlement periods $(d^* + 1\ to\ d^* + n_D^+)$ provide an indication of the future renewable generation forecast for efficient resource management.

The last three time horizons are considered by the optimal dispatch unit (Figure 9.1) of the DHRB.

9.1.3 OPTIMIZATION PROBLEM DEFINITION

This chapter deals with the optimal dispatch of DHRB power for injecting into the grid from the perspective of the DHRB operator. Considering the past announced, target, and future expected settlement periods, an optimization problem can be

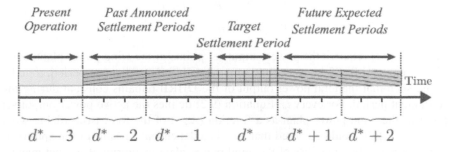

FIGURE 9.2 DHRB dispatch time horizons.

formulated to find the optimal set points of a DHRB. The constraints of the optimization problem are given by Equations (9.1) to (9.14).

$$P_{h,d} = \sum_{s=1}^{S} P_{s,t_d} + \sum_{w=1}^{W} P_{w,t_d} + \sum_{b=1}^{B} P_{b,t_d} \quad \forall \; t_d \in T_d \tag{9.1}$$

$$P_{w,t_d} = \lambda_{w,t_d} P'_{w,t_d} \tag{9.2}$$

$$\omega_{w,t_d} \lambda_w^{min} \leq \lambda_{w,t_d} \leq \omega_{w,t_d} \lambda_w^{max} \tag{9.3}$$

$$P_{s,t_d} = \lambda_{s,t_d} P'_{s,t_d} \tag{9.4}$$

$$\omega_{s,t_d} \lambda_s^{min} \leq \lambda_{s,t_d} \leq \omega_{s,t_d} \lambda_s^{max} \tag{9.5}$$

$$P_{b,t_d} = \lambda_{b,dc,t_d} P_{b,dc,t_d}^{max} \delta_{b,dc} - \lambda_{b,c,t_d} P_{b,c,t_d}^{max} \delta_{b,c}^{-1} \tag{9.6}$$

$$\omega_{b,dc,t_d} \lambda_{b,dc}^{min} \kappa_{b,dc,t_d}^{min} \leq \lambda_{b,dc,t_d} \leq \omega_{b,dc,t_d} \lambda_{b,dc}^{max} \kappa_{b,dc,t_d}^{max} \tag{9.7}$$

$$\omega_{b,c,t_d} \lambda_{b,c}^{min} \kappa_{b,c,t_d}^{min} \leq \lambda_{b,c,t_d} \leq \omega_{b,c,t_d} \lambda_{b,c}^{max} \kappa_{b,c,t_d}^{max} \tag{9.8}$$

$$\omega_{b,c,t_d} + \omega_{b,dc,t_d} \leq 1 \tag{9.9}$$

$$E_{b,t_d} N_D N_T = (N_D N_T - \delta_{b,e}) \quad E_{b,t_d-1} + \lambda_{b,c,t_d} P_{b,c}^{max} - \lambda_{b,dc,t_d} P_{b,dc}^{max} \tag{9.10}$$

$$E_b^{max} \lambda_{b,e}^{min} \kappa_{b,e,t_d}^{min} \leq E_{b,t_d} \leq E_b^{max} \lambda_{b,e}^{max} \kappa_{b,e,t_d}^{max} \tag{9.11}$$

$$\sum_{b=1}^{B} -E_b^{max} \lambda_{b,e}^{min} \kappa_{b,e,t_d}^{min} \delta_{b,e} \delta_{b,c}^{-1} \leq P_{b,d} \quad \forall \; t_d \in T_d \tag{9.12}$$

$$\lambda_{b,c,t_d} P_{b,c}^{max} \leq \sum_{s=1}^{S} P_{s,t_d} + \sum_{w=1}^{W} P_{w,t_d} + E_b^{max} \lambda_{b,e}^{min} \kappa_{b,e,t_d}^{min} \delta_{b,e} \delta_{b,c}^{-1} \tag{9.13}$$

$$\begin{cases} P_{b,d}^e = P_{b,d} - \overline{P}_{b,d} & d < d^* \\ P_{b,d}^e = 0 \; d \geq d^* \end{cases} \tag{9.14}$$

Equation (9.1) requires the power output set points of the wind power plant, solar PV system, and BESS in every time step, t_d, within the settlement period, d, to be found such that the total power output of the DHRB remains constant during the settlement period. For every t_d, Equation (9.2) defines the active power output set point of the wind power plant as a fraction (λ_{w,t_d}) of the forecast available wind power (P'_{w,t_d}). It should be noted that λ_{w,t_d} is a dependent variable of the optimization problem. The capability of the wind power plant in altering its output may be narrower than the theoretical range, i.e., zero to the maximum available power; this constraint is imposed by Equation (9.3). The binary variable, ω_{w,t_d}, enables the wind power plant to be turned off, i.e., $P_{w,t_d} = 0$ in time steps that $\lambda_w^{min} \leq \lambda_{w,t_d}$ is not feasible. Similar to the wind power plant, the active power output set point of the solar PV system for every t_d is defined by Equation (9.4) and its utilization factor is constrained by Equation (9.5). The power output of BESS at each t_d is given by Equation (9.6). The first term is the power injected due to the BESS discharging, and the second term is the power drawn for charging it. BESS is discharging when $P_{b,t_d} > 0$ and charging when $P_{b,t_d} < 0$. As noted, the discharging and charging power are defined as fractions (λ_{b,dc,t_d} and λ_{b,c,t_d}) of the maximum discharge and charge rate of BESS (P_{b,dc,t_d}^{max} and P_{b,c,t_d}^{max}) subject to its discharge and charge efficiency ($\delta_{b,dc}$ and $\delta_{b,c}$). Due to technical and physical limits, the feasible/recommended range for discharging and charging of the BESS may be narrower than zero to the maximum rate. This constraint is included by Equations (9.7) and (9.8). The degradation factors, κ_{b,dc,t_d}^{max}, κ_{b,dc,t_d}^{min}, κ_{b,c,t_d}^{max}, and κ_{b,c,t_d}^{min} reflect the power fading of BESS, and their derivation is discussed in Section 9.1.4. The role of binary variables, ω_{b,dc,t_d} and ω_{b,c,t_d}, is similar to that explained for the wind power plant and solar PV system. It is evident that simultaneous discharging and charging of BESS is not a viable and/or feasible solution, therefore, Equation (9.9) allows BESS to only discharge or charge at every t_d by requiring $\omega_{b,dc,t_d} = 0$ and/or $\omega_{b,dc,t_d} = 0$. The stored energy in the BESS at the end of every time step (E_{b,t_d}) is dependent on the stored energy at the end of the previous time step (E_{b,t_d-1}) as well as the power injected/drawn by the BESS during the current time step; this is given by Equation (9.10). The hourly self-discharge of BESS (at the rate of $\delta_{b,e}$) is mimicked by the first term in Equation (9.10). To prolong the battery lifetime, the state of the charge is maintained within $[\lambda_{b,e}^{min} \lambda_{b,e}^{max}]$ per Equation (9.11) in all time steps. The degradation factors κ_{b,e,t_d}^{max} and κ_{b,e,t_d}^{min} reflect the capacity fading of BESS, and their derivation is discussed in Section 9.1.4. There may be time steps at which the stored energy in BESS is at its minimum limit. If, due to self-discharge, the stored energy in BESS further drops, there will be breaching of the BESS operation constraints. Hence, the combination of Equation (9.11) and the self-discharge component of BESS (i.e., $\delta_{b,e}$) can make the problem infeasible. Accordingly, Equation (9.12) allows the DHRB output to become negative, i.e., consume a relatively small amount of power (e.g., in Li-ion batteries up to 0.0042–0.0125% per hour [4] of the minimum allowed energy level subject to charge efficiency). This enables canceling the BESS self-discharge component when enough renewable generation is not available. It is trivial that the frequency of such a power draw from the grid can be reduced by increasing the forecast horizon; however, it is not guaranteed to be avoided. Further, Equation (9.13) limits the BESS charge power at every time step to the total renewable power output plus the least power required to cancel the BESS self-discharge component. This constraint implies that BESS is only used for the provision of dispatchable renewable power and not energy arbitrage. However, if the DHRB operator decides otherwise, Equations (9.12) and (9.13) need to be amended accordingly. Equation (9.14) defines the power shortage as the difference between the DHRB set point and the committed power for past announced settlement periods. This value is equal to 0 for the target and future expected settlement periods since the DHRB has not been committed for these settlement periods yet. For the purpose of comparison, the constraints given by

Equations (9.1) to (9.14) are used to define two separate optimization problems, each corresponding to an operation strategy.

- **Maximization of DHRB Revenues**

$$Max \sum_{d=d^*-n_{\bar{D}}}^{d^*+n_{\bar{D}}^+} \left(\frac{P_{h,d}C_{h,d} - C_{h,d}^P P_{h,d}^e}{N_D} - \sum_{b=1}^{B} \sum_{t_d \in T_d} \frac{C_b \chi_{b,t_d}}{\chi_{b,total}} \right) \tag{9.15}$$

subject to Equations (9.1) to (9.14).

In the objective function, Equation (9.15), the predicted electricity price at every d ($C_{h,d}$) is used to calculate the income of the DHRB subject to penalty on $P_{h,d}^e$. Excessive use of BESS shortens its lifetime. To avoid exhausting BESS (to further harvest renewables), the latter term in Equation (9.15) applies a penalty to BESS energy transactions (χ_{b,t_d}). The penalty factor is based on the BESS capital cost (C_b) and the BESS nominal energy transaction ($\chi_{b,total}$). BESS energy transaction, χ_{b,t_d}, is defined as the summation of all energy input and output to BESS due to its charging, discharging and self-discharge during the time step of interest, t_d. Similarly, BESS nominal energy transaction, $\chi_{b,total}$, is the total amount of energy that can be charged to and discharged from BESS during its useful lifetime, i.e., before BESS' capacity falls below a certain percentage (EOL_b) of its rated capacity. Accordingly, χ_{b,t_d} and $\chi_{b,total}$ are given by Equations (9.16) and (9.17), respectively. It is worth mentioning that Equation (9.17) is the opened form of a similar equation given in [5].

$$\chi_{b,t_d} = \frac{E_{b,t_d-1}\delta_{b,e} + \lambda_{b,c,t_d}P_{b,c}^{max} + \lambda_{b,dc,t_d}P_{b,dc}^{max}}{N_D N_T} \tag{9.16}$$

$$\chi_{b,total} = \frac{\left(\lambda_{b,e}^{max} - \lambda_{b,e}^{min}\right)\delta_{b,c}\delta_{b,dc}E_b^{max}\left(EOL_b+1\right)}{\left(N_b+1\right)^{-1}} \tag{9.17}$$

- **Maximization of Renewable Harvesting**

$$Max \sum_{d=d^*-n_{\bar{D}}}^{d^*+n_{\bar{D}}^+} \sum_{t_d \in T_d} \frac{1}{N_D N_T} \left(\sum_{s=1}^{S} P_{s,t_d} + \sum_{w=1}^{W} P_{w,t_d} - P_{h,d}^e M \right) \tag{9.18}$$

subject to Equations (9.1) to (9.14),
where M is a large number to ensure $P_{h,d}^e = 0$, if the problem doesn't become infeasible. The two defined mixed integer linear problems (MILPs) may be solved (separately) for a given number of past announced settlement periods and future expected settlement periods to find the power set points for the wind power plant (P_{w,t_d}), solar PV system (P_{s,t_d}), and BESS (P_{b,t_d}) at every time step such that the respective objective is realized while the power output of the DHRB remains constant during each settlement period and the results shall be compared.

9.1.4 BATTERY DEGRADATION

Battery modules in a BESS degrade with time, utilization, and environmental and operating conditions. To avoid overestimation of the available resources, it is essential to consider this factor for long-term BESS operation and sizing. Several degradation models have been proposed for various battery technologies [6]. For this chapter the linear model presented by [5] was employed. Based on this model the battery degradation coefficients are defined.

$$\kappa_{b,e,td}^{max} = \kappa_{b,e,td-1}^{max} + (EOL_b - 1)\chi_{b,td-1}/\chi_{b,total} \qquad (9.19)$$

$$\kappa_{b,e,td}^{min} = \kappa_{b,e,td-1}^{min} + (EOL_b - 1)\chi_{b,td-1}/\chi_{b,total} \qquad (9.20)$$

$$\kappa_{b,c,td}^{max} = \kappa_{b,dc,td}^{max} = \kappa_{b,c,td}^{min} = \kappa_{b,dc,td}^{min} = 1 \qquad (9.21)$$

It is emphasized that more sophisticated battery degradation models may also be incorporated [7–10]. Direct inclusion of the battery degradation model given by Equations (9.19) to (9.21), i.e., treating degradation coefficients as variables, does not affect the MILP nature of the optimization problems defined in Section 9.1.3. However, to enable the integration of more sophisticated degradation models, the battery degradation coefficients shall be calculated prior to solving the optimization problem and treated as parameters. To this end, prior to solving the optimization problem, the battery degradation coefficients are calculated for two time steps:

- **Start of the first settlement period:** This is the beginning of $d^* - n_D^-$ (based on previous operating points) and is named κ_{start}.
- **End of the last settlement period:** This is the end of $d^* + n_D^+$ (based on worst operating scenario, i.e., $\omega_{b,c,td}\lambda_{b,c,td} + \omega_{b,dc,td}\lambda_{b,c,td} = 1$) and is named κ_{end}.

Provided that the time length of $d^* - n_D^-$ to $d^* + n_D^+$ is sufficiently small, a linear approximation may be used to calculate the degradation coefficients at every time step, using κ_{start} and κ_{end}:

$$\kappa_{td} = \frac{(\kappa_{end} - \kappa_{start})\left(d - d^* + n_D^- + frac\left(\frac{t_d - 1}{N_T}\right)\right)}{n_D^+ + n_D^- + 1} + \kappa_{start} \qquad (9.22)$$

In Equation (9.22), the function $frac(.)$ yields the fractional part (decimal part) of the input value. The degradation coefficients found with this approach are then passed to the optimization problem as parameters (rather than variables). It is worth mentioning that compared with BESS, the degradation of wind power plants and solar PV systems is a slower process [11, 12]. It can be integrated in the aggregation unit for estimation of $P'_{w,td}$ and $P'_{s,td}$ at every time step. It should be noted that, the units for N_D and N_T are $\frac{settlement\ period}{hour}$ and $\frac{time\ step}{settlement\ period}$, respectively; they are used in

Equations (9.10), (9.15), (9.16), (9.18), and (9.22) to take into account the length of the considered time interval in each equation.

9.1.5 ROLLING ALGORITHM

To adapt the presented methodology for the continuous operation of the DHRB, the rolling algorithm shall be employed per Figure 9.3. Rolling is a technique for optimization over a time series. It facilitates continuous optimization using updated data (e.g., forecast, battery degradation coefficients) in every run [13]. In this algorithm, the optimization problem is solved for the $d^* - n_D^-$ to $d^* + n_D^+$ time window and results are stored. The target settlement period is increased by 1, i.e., the time window slides ahead by one settlement period. Once forecast data are available for $d^* + n_D^+$, the optimization problem is solved again to find set points for $d^* + n_D^+$ and update set points for the other settlement periods. As implied, the forecast for every t_d can be updated (and perfected) $n_D^+ + n_D^- + 1$ times before d leaves the sliding time window. Employing the rolling algorithm enables readjustment of the solar-wind and BESS set points according to the most recent forecast not only for the target but also the past announced settlement periods as long as the settlement period is still within the $d^* - n_D^-$ to $d^* + n_D^+$ time window. In this chapter, BESS sizing is considered. Therefore, the stop criterion is the BESS end of life; however, any other stop criteria may be used by the DHRB operator.

It should be noted that the distribution of renewables forecast error depends on the forecast horizon. The nearer in time the forecast subject is, the more concentrated the error around the mean value [14, 15]. The accuracy of forecast reduces with the increase in forecast horizon. For a given confidence level, near-term forecasting techniques can provide a narrower confidence interval compared with day-ahead

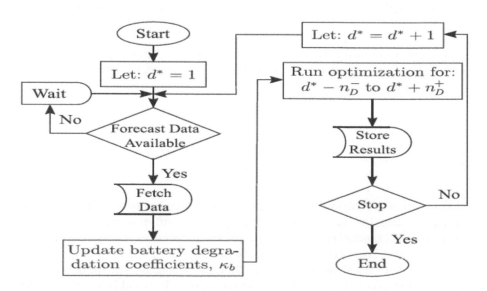

FIGURE 9.3 Rolling algorithm for DHRB.

forecasting. Therefore, this methodology is not suitable if the target settlement period is several hours ahead of present operation. As implied, instead of the conventional day-ahead market, this methodology is suitable in markets such as the intraday continuous market.

The presented methodology was implemented using Python scripting language and Microsoft SQL Server 2016. The optimization problem was handled by Pyomo [16] and solved using the Cplex MILP solver.

9.2 CASE STUDY

To demonstrate the applicability of the presented methodology, an 80-MW DHRB consisting of a 50-MW wind power plant, a 30-MW solar PV system, and a 5 MWh BESS (NMC type [17]) is used. 10-minute resolution wind and solar power generation profiles were obtained from [18]. These were assumed to correspond to the lower confidence limit of forecast with a sufficiently high confidence level, e.g., 95%. However, for real-world operation, the renewables forecast is generated in real time by the forecast and aggregation units (see Figure 9.1) based on the lower confidence limit with the desired confidence level for several time steps ahead. It was assumed that each settlement period is 30 minutes long and the electricity market operator requires power bids to be submitted 1 hour ahead (these can be adjusted based on market requirements in each power system). For efficient management of BESS, two future expected settlement periods were used; however, more can be considered depending on the availability of forecast data and computational cost. It was assumed that the wind power plant and solar PV system can regulate their output between 0 and maximum available power at every time step. The DHRB injects power under a contract-for-difference (CFD) scheme [19] with £60/MWh strike price (2019 £) [20]. The short-term balancing and reliability cost increase due to the intermittency of renewables is assumed to be £5/MWh [21, 22]. Because a DHRB guarantees the dispatch, the additional short-term balancing and reliability cost can be avoided and deemed as a potential income for the DHRB. Wind power plant and PV system parameters, BESS parameters, and other parameters of the optimization problem are summarized in Tables 9.1 to 9.3, respectively. It is acknowledged that in a market that exposes renewable generation to the market electricity prices, a time series of predicted $C_{h,d}$ should be used instead. This requires prediction of $C_{h,d}$ based on the historic values, loading and fuel prices, among others, which is beyond the scope of this chapter. Moreover, the committed/dispatched power depends on the market

---

TABLE 9.1

Wind Power Plant and PV System Parameters

Parameter	Value	Parameter	Value
Max $\left(P'_{w,t_d}\right)$	50 MW	Max $\left(P'_{s,t_d}\right)$	30 MW
λ_w^{max}	1.0	λ_s^{max}	1.0
λ_w^{min}	0.0	λ_s^{min}	0.0

TABLE 9.2

Battery Parameters

Parameter	Value	Parameter	Value
E_b^{max}	5 MWh	$\lambda_{b,c}^{max}$	1.0
$\delta_{b,e}$	0.005%	$\lambda_{b,c}^{mim}$	0.0
$\lambda_{b,e}^{max}$	1.0	$P_{b,dc}^{max}$	10 MW
$\lambda_{b,e}^{mim}$	0.2	$\delta_{b,dc}$	0.95
$P_{b,c}^{max}$	10 MW	$\lambda_{b,dc}^{max}$	1.0
$\delta_{b,c}$	0.95	$\lambda_{b,dc}^{mim}$	0.0
EOL_b	0.65	N_b	4,000
C_b	£261/kWh	-	-

TABLE 9.3

Other Parameters

Parameter	Value	Parameter	Value
$C_{h,d}$	£65/MWh	$C_{h,d}^e$	£2,700/MWh
N_D	2	n_D^+	2
N_T	3	n_D^-	2

and the power system operator. The hardest situation for the DHRB is when all the dispatchable power is committed/dispatched; this is the assumption made for $\overline{P}_{h,d}$.

An Intel Xeon E3 1240 v5 CPU, 16 GB DDR4 2.4 GHz memory, and 100-Mbps connection to the database were used in the case study. Fetching data, updating degradation coefficients, solving the optimization problem for $d^* - n_D^-$ to $d^* + n_D^+$, and storing the results took on average 2 seconds for every iteration (every target settlement period).

9.3 RESULTS AND DISCUSSION

9.3.1 OPERATION STRATEGY

The presented rolling algorithm was run to dispatch the DHRB for 1 year with two different strategies, maximization of DHRB revenues (max revenue) and maximization of renewable harvesting (max harvest). Power output set points were optimized by the algorithm for the wind power plant, solar PV system, and BESS. Figures 9.4 to 9.6 illustrate the power set points for the wind power plant, solar PV systems, and BESS, respectively, for a sample summer day. When considered individually, the power output of the wind power plant, solar PV system, and BESS fluctuate significantly within every settlement period (30 minutes). However, the net power produced by the DHRB stays constant for the duration of every settlement period; this is shown in Figure 9.7.

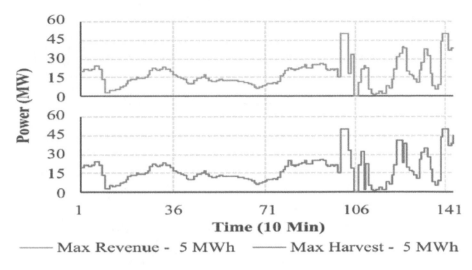

FIGURE 9.4 Wind power set point on a summer day.

It is evident from these figures that the power set points found by max revenue and max harvest strategies follow a similar trend; however, in some time steps they are not identical. This is more visible in the BESS set point. This discrepancy is mainly due to the difference in charging/discharging BESS (and consequently charge state) in the studied strategies. Accordingly, the use of BESS is categorized:

- **Inevitable use:** Because any uncommitted power ($P^e_{h,d}$) is heavily penalized in both strategies (by $C^P_{h,d}$ in max revenue and big M in max harvest), BESS is used to meet the committed power for $d^* - n_{\bar{D}}$, $\overline{P}_{h,d^*-n_{\bar{D}}}$.
- **Optimum use:** Considering that a penalty is applied to the energy transactions of BESS (charge/discharge) in Equation (9.15), max revenue entails

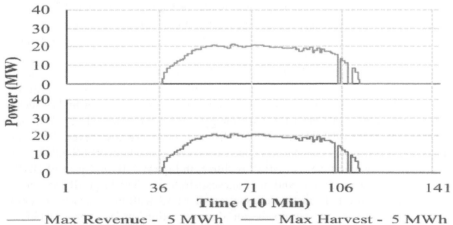

FIGURE 9.5 Solar power set point on a summer day.

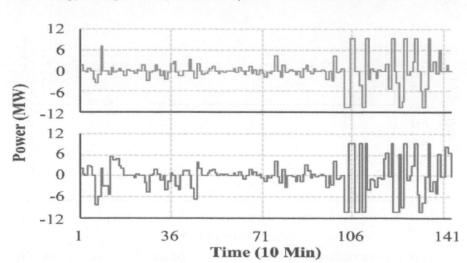

FIGURE 9.6 BESS power set point on a summer day (negative values correspond to BESS charging).

minimal use of BESS unless it's essential (inevitable use) or it results in a higher revenue in the d^* to $d^* + n_D^+$ interval. In contrast, the use of BESS is not penalized in Equation (9.18). Accordingly, the optimization problem uses BESS as needed to avoid nondispatchable power as much as possible, i.e., maximize harvesting of renewables. It should be noted that nondispatchable power is the portion of power that is neither dispatchable nor stored in BESS.

The BESS state of charge for the sample summer day is shown in Figure 9.8. It is noted that the energy input and output for the max revenue strategy is smaller than that of the max harvest. A similar trend holds for the rest of the days of the year. Figure 9.9 illustrates the annual cumulative probability of the BESS power set point. The set point is within 0.5 MW in 57% and 45% of the time steps for max revenue and max harvest, respectively.

As implied by Figure 9.7, part of the available renewable energy is nondispatchable. Table 9.4 lists the annual dispatchable and nondispatchable energy achieved in the studied strategies. Compared with max revenue, at the cost of higher energy transaction in BESS, the max harvest strategy yielded approximately 1 GWh more dispatchable energy by reducing the nondispatchable energy. However, the excess usage of BESS in the max harvest case raises concerns about BESS exhaustion. The degradation coefficient of BESS is shown in Figure 9.10 for 1 year of DHRB operation with max revenue and max harvest strategies. The larger BESS energy transaction manifests in faster degradation of BESS with max harvest strategy. Therefore, max harvest may not be an ideal operation strategy for a DHRB. It is worth mentioning that indeed max revenue and max harvest can be combined as a multi-objective problem by applying weighting factors and summing up

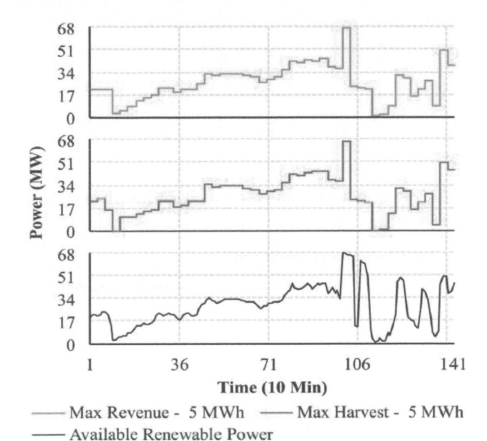

FIGURE 9.7   DHRB output on a summer day.

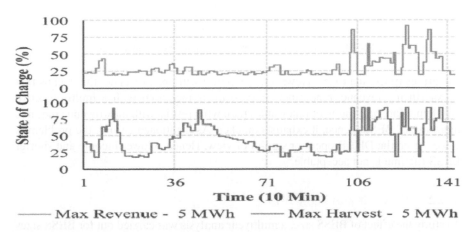

FIGURE 9.8   BESS state of charge on a summer day.

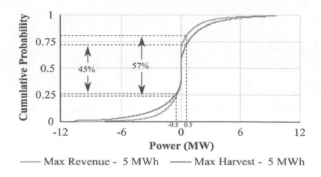

FIGURE 9.9 Annual cumulative probability of BESS power set point.

TABLE 9.4

DHRB Annual Energy Constitution

Strategy	Dispatchable Energy (GWh)	Nondispatchable Energy (GWh)
Max revenue	257.75	1.41
Max harvest	258.71	0.38

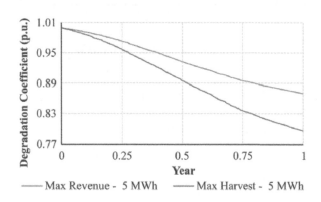

FIGURE 9.10 BESS degradation coefficient for the first year of operation.

Equations (9.15) and (9.18). However, compared with max revenue, this means that by sacrificing revenue, renewables harvesting is increased with a multi-objective problem. More harvesting of renewables than what results with max revenue requires more use of BESS, hence, it exhausts quicker. As mentioned earlier, the problem is solved from the DHRB operator's point of view. Hence, a lower income and shorter BESS lifetime is not acceptable.

9.3.2 BESS Size and Economy

To study the effect of BESS size, a multiyear analysis was carried out for BESS sizes ranging from 2.5 to 17.5 MWh, with 2.5-MWh increments (rated at 2C) using the max revenue strategy. For this purpose, the 1-year wind and solar generation profile

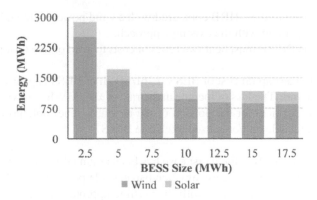

FIGURE 9.11 Annual nondispatchable energy at various BESS sizes for the revenue maximization scenario.

was duplicated for 10 years and used as an input to the presented algorithm. The total nondispatchable energy in the first year of operation of DHRB is depicted in Figure 9.11. Initially, the annual nondispatchable energy reduced with the increase in BESS size; however, after a certain size of BESS (i.e., 10 MWh) the improvement was marginal. Figure 9.12 illustrates the degradation coefficient of BESS through its lifetime (note that EOL = 0.65). The increase in the BESS size prolonged its lifetime because the larger the BESS size, the smaller the per-unit depth of charge and discharge is. The existing approach for wind and solar generation is to inject all the generated power to the grid in a nondispatchable fashion (if wind curtailment is not mandated by the system operator). Provision of sufficient flexible reserve is a key to tackle the intermittency of the injected power. Hence, additional balancing and reliability costs will be inflicted due to the integration of nondispatchable renewable generation. In the case study, this cost is assumed as £5/MWh [21, 22] for the renewable energy injected to the grid. Because a DHRB guarantees the dispatch, it does not inflict such an additional cost. Therefore, this amount is taken as the only additional reward/

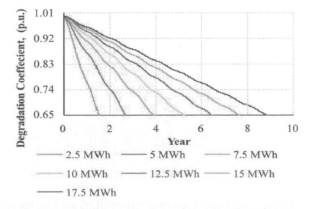

FIGURE 9.12 BESS degradation coefficient at various BESS sizes for the revenue maximization scenario.

incentive provided to the DHRB compared with a wind power plant and/or solar PV system that is operated with the existing approach.

However, the DHRB must bear two types of costs to avail of this reward/incentive,

- **Capital:** The capital cost of BESS (£261/MWh of BESS capacity.
- **Operational:** The loss of income at the rate of £60/MWh (equivalent to the CFD strike price) for the portion of the available renewable power that is nondispatchable and/or is used to charge BESS

It is worth mentioning that the worst case has been assumed for the operational cost in the case study. In the best case, any nondispatchable power from DHRB shall be injected into the grid and treated with the existing approach; hence, the income for this portion of the available power is secured.

An economic analysis was carried out on the simulation results for the BESS sizes and DHRB operation strategy of interest, i.e., max revenue. The additional income/loss of the DHRB compared with a similar size wind power plant and solar PV system operated with the existing approach was calculated for each case. This was used to establish the average annual rate of return for the extra investment on BESS that the DHRB should incur. As seen in Figure 9.13, for all of the BESS sizes, the rate of return is a positive value, meaning that despite the operational cost borne, because of the £5/MWh incentive/reward, the DHRB is able to produce higher revenue than a similar size wind power plant and solar PV system operated with the existing approach. Moreover, it is evident from Figure 9.13 that the lower the BESS size is the better the annual average rate. However, it should be considered that the lifetime of the BESS will also be shorter. This means that the DHRB will have less time to make revenue before it needs to replace batteries. It is pointed out that the nonlinearity of the annual rate of return against the BESS capacity is because the total capital cost increases with BESS size; however, the average annual additional income of DHRB (compared with a similar size wind power plant and solar PV system operated with the existing approach) does not increase substantially after 10 MWh. Figure 9.14 shows the cumulative additional cash flow of the DHRB when operated with max revenue strategy (compared with a similar size wind power plant

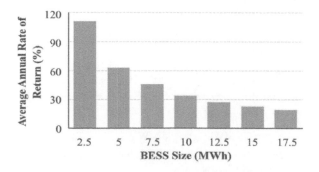

FIGURE 9.13 Average annual rate of return at various BESS sizes for the revenue maximization scenario.

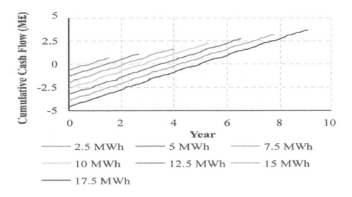

FIGURE 9.14 Cumulative cash flow at various BESS sizes for the revenue maximization scenario.

and solar PV system operated with the existing approach) through the lifetime of the BESS for various sizes of BESS. To provide further ground for comparison, the same analysis was also carried out for DHRBs operated with the max harvest strategy; the results are shown in Figure 9.15. It is noted that with both DHRB operation strategies, max revenue and max harvest, regardless of the BESS size, the slope of the cumulative cash flow trend is positive. This indicates that DHRB is able produce higher revenue compared with a similar size wind power plant and solar PV system operated with the existing approach. In all cases, it is seen that the DHRB is able to recover (cumulative cash flow becomes positive) its additional investment (on BESS) solely through the additional £5/MWh incentive/reward. Hence, DHRB is superior to existing approaches. Moreover, in general, compared with max harvest, the cumulative cash flow of DHRB at the end of life of the BESS is higher when DHRB is operated with max revenue strategy. The higher the size of the BESS, the higher the cumulative cash at the end of life of BESS will be when max revenue strategy is used. Both phenomena occur because the longer the BESS lifetime, the more dispatchable energy can be injected by the DHRB. This denotes the importance of the operation strategy and the right choice of BESS size.

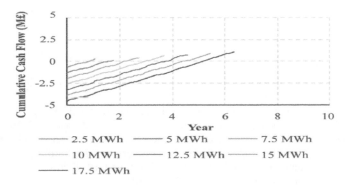

FIGURE 9.15 Cumulative cash flow at various BESS sizes for the renewable harvesting maximization scenario.

REFERENCES

[1] Al-Zadjali, S., et al.: An accurate, light-weight wind speed predictor for renewable energy management systems. Energies 12(22), 1–20 (2019)

[2] Dobbs, A. et al.: Short-term solar forecasting performance of popular machine learning algorithms. In: 7th International Workshop on the Integration of Solar Power into Power Systems, pp. 1–6. National Renewable Energy Laboratory, Golden, CO (2017)

[3] Elexon: The Electricity Trading Arrangements: A Beginner's Guide. Elexon Ltd. (2018). https://www.elexon.co.uk/guidancenote/beginnersguide/. Accessed 13 Aug 2018

[4] Chatzivasileiadi, A., Ampatzi, E., Knight, I.: Characteristics of electrical energy storage technologies and their applications in buildings. Renewable Sustainable Energy Rev. 25, 814–830 (2013)

[5] Pourmousavi Kani, S.A., Wild, P., Saha, T.K.: Improving predictability of renewable generation through optimal battery sizing. IEEE Trans. Sustain. Energy 11(1), 37–47 (2020)

[6] Ramadesigan, V., et al.: Modeling and simulation of lithium-ion batteries from a systems engineering perspective. J. Electrochem. Soc. 159(3), R31–R45 (2012)

[7] Tourani, A., White, P., Ivey, P.: A multi scale multi-dimensional thermos electrochemical modelling of high capacity lithium-ion cells. J. Power Sources 255, 360–367 (2014)

[8] Garofalini, S.H.: Molecular dynamics simulations of Li transport between cathode crystals. J. Power Sources 110(2), 412–415 (2002)

[9] Shen, J., Dusmez, S., Khaligh, A.: Optimization of sizing and battery cycle life in battery/ultracapacitor hybrid energy storage systems for electric vehicle applications. IEEE Trans. Ind. Inf. 10(4), 2112–2121 (2014)

[10] Song, J., et al.: Development of a Markov-chain-based energy storage model for power supply availability assessment of photovoltaic generation plants. IEEE Trans. Sustain. Energy 4(2), 491–500 (2013)

[11] Staffell, I., Green, R.: How does wind farm performance decline with age? Renewable Energy 66, 775–786 (2014)

[12] Sun, X., Chavali, R.V.K., Alam, M.A.: Real-time monitoring and diagnosis of photovoltaic system degradation only using maximum power point–the Suns-Vmp method. Prog. Photovoltaics Res. Appl. 27(1), 55–66 (2019)

[13] O'Connell, A., Flynn, D., Keane, A.: Rolling multi-period optimization to control electric vehicle charging in distribution networks. IEEE Trans. Power Syst. 29(1), 340–348 (2014)

[14] Madsen, H., et al.: Standardizing the performance evaluation of short-term wind power prediction models. Wind Eng. 29(6), 475–489 (2005)

[15] Sharma, V., Chandel, S.S.: Performance and degradation analysis for long term reliability of solar photovoltaic systems: A review. Renewable Sustain. Energy Rev. 27, 753–767 (2013)

[16] Hart, W.E. et al.: Pyomo–optimization modeling in python. vol. 67. 2nd ed. Springer, Berlin, Heidelberg (2017)

[17] Kokam Energy Sotage Solutions. http://kokam.com. Accessed 13 Aug 2018

[18] Grid Data and Tools. https://nrel.gov/grid/data-tools.html. Accessed 13 Aug 2018

[19] UK Government: FiT contract for difference standard terms and conditions. (2017). https://assets.publishing.service.gov.uk/government/uploads/system/uploads/attachment_data/file/599098/FINAL_CFD_Standard_Terms_and_Conditions_V2-_13_March_2017_.pdf. Accessed 13 Aug 2018

[20] Arup: Market stabilization analysis: Enabling investment in established low carbon electricity generation. Arup Group Limited, (2017). https://www.arup.com/-/media/arup/files/publications/e/enablinginvestment-in-established-low-carbon-electricity-generation.pdf. Accessed 13 Aug 2018
[21] Heptonstall, P., Gross, R., Steiner, F.: The costs and impacts of intermittency–2016 update. UK Energy Research Centre, London (2017)
[22] Gross, R., et al.: Renewables and the grid: understanding intermittency. Proc. Inst. Civ. Eng. 160(1), 31–41 (2007)

10 Real-Time Control in Hybrid Renewable Energy Systems and Validation Using RTDS

Mohammed Al-Busaidi, Amer Al-Hinai,
Hassan Haes Alhelou, Rashid Al-Abri

CONTENTS

DOI: 10.1201/9781003307433-10

10.1 DEVELOPING HYBRID POWER PLANT IN RTDS

The hybrid power plant and renewable energy sources models in this project were simulated using the real-time digital simulator (RTDS). An RTDS simulator consists of hardware and software packages to carry out a real-time power system simulation for hardware-in-the-loop (HIL) testing of protection, control, and power components. Essentially, RTDS is a globally acknowledged standard real-time power system simulator utilized by more than 400 customers distributed in 43 countries around the world [1], thus, making it trusted worldwide by leading power systems manufacturers, utilities, and research institutions. It was developed by Manitoba HVDC Research Centre in cooperation with Manitoba Hydro in Canada.

The hybrid power plant is modeled in RTDS using preexisting validated and industry-approved models of photovoltaic (PV), wind turbines, battery storage systems, and the grid model. These models have been used in many studies. A list containing some of the studies is mentioned in [1, 2].

The hybrid power plant consists of PV, a doubly-fed induction generator (DFIG), and battery storage systems. The capacity ratios of these energy sources are derived from the results of [3]. It is concluded that the optimal PV size would be 50% of the total DFIG rating. As for the battery storage, the capacity should not exceed 6% of the total hybrid power plant rating provided a fast storage system is integrated with the system to eliminate the effect of fast fluctuations from wind turbines.

However, a major factor affecting the total number of modeled equipment is the available computing sources of RTDS in Sultan Qaboos University (SQU). Currently, only three cores are licensed in RTDS, which limits the size of the hybrid power plant to only 10 MW, as detailed in Table 10.1. It is worth mentioning that the main objective of the simulations is to examine the performance of the designed control scheme rather than to focus on the hybrid plant itself. Hence, the size of the plant is not a primary variable in the problem in study. The designed controller scheme should be capable of dealing with any plant size.

The overall structure of the hybrid plant model is demonstrated in Figure 10.1.

The specifications of the PV module and DFIG are shown in Tables 10.2 to Table 10.4. The PV module I–V characteristics curve is displayed in Figure 10.2. Figure 10.3 illustrates the characteristics of DFIG.

TABLE 10.1

Hybrid Power Plant Energy Sources Capacity

Energy Source	Single Source Rating	Number of Equipment	Total Rating
DFIG	2 MW	3	6 MW
PV string	1 MW	3	3 MW
Battery system	1 MW	1	0.65 MWh

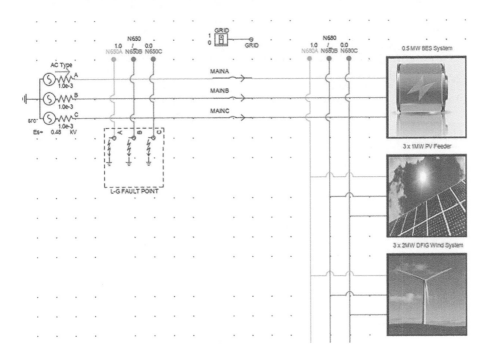

FIGURE 10.1   Overview of the hybrid power plant model in RTDS.

In Table 10.4, the parameters C1, C2, C3, C4, C5, and C6refer to the coefficients in Equation (10.1).

$$C_p(\beta,\lambda) = c_1\left(\frac{c_2}{\lambda_i} - c_3\beta - c_4\right)e^{\frac{-c_5}{\lambda_i}} + c_6\lambda \tag{10.1}$$

TABLE 10.2

PV Module Specification at Standard Test Conditions (Irradiance 1000 W/m² and 25°C)

Specification	Unit	Value
Maximum power (P_{max})	W	385
MPP voltage (V_{mpp})	V	41
MPP current (I_{mpp})	A	9.40
Open circuit voltage (V_{oc})	V	49.1
Short circuit current (I_{sc})	A	10.11
Module efficiency [%]		18.2
Maximum system voltage	V	1000
Cells		6×12
Cell type		Monocrystalline/N-type

TABLE 10.3
DFIG Specification

Specification	Value
Rated MVA	2.2 MW
Maximum power	2.0 MW
Stator voltage (L–L)	0.69 kV
Stator resistance	0.001 Ω
Rotor resistance	0.0013 Ω
Stator reactance	0.963 Ω
Magnetizing reactance	0.941 Ω
Rotor reactance	0.965 Ω
Rotor/stator turns ratio	2.6377
Inertia constant	1.5 MWs/MVA
Rated wind speed	12 m/s
Cut-in wind speed	6 m/s

TABLE 10.4
DFIG Model Coefficient Values

Name	Description	Value
C_1	Cp (lmda,Beta) = (n1 – n2*Beta)*	0.47
C_2	sin(1.57*(lmda – 3y)/(n3 – n4*Beta))	0.0167
C_3	–(lmda – 3y)*n5*Beta + n6/(1 + lmda)	7.5
C_4		0.15
C_5	Y = 1 – exp(–lmda/3)	0.00184
C_6		0.01

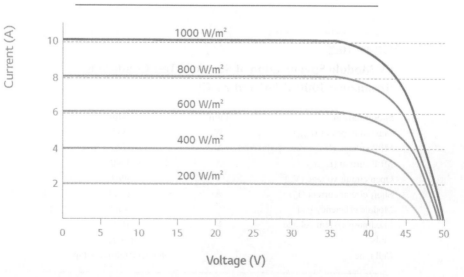

FIGURE 10.2 PV module I–V characteristics curve.

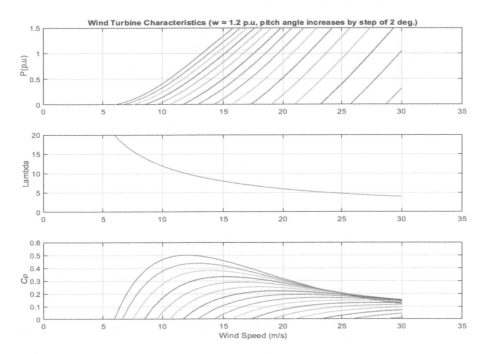

FIGURE 10.3 DFIG characteristics curves while incrementing pitch angle: (a) P(p.u), (b) lambda (λ), (c) Cp (λ, β).

where C_p is the performance coefficient, which is a function of blade pitch angle β and tip speed ratio λ [4].

10.2 MODELING LOCAL INVERTER CONTROLLERS IN RTDS

The upcoming subsections detail the development carried to model the renewable energy sources primary controllers. Those controllers are attached to every power converter. Every primary controller is integrated with multiple operation modes as discussed in this section.

10.2.1 DFIG ROTOR-SIDE CONTROL

The DFIG rotor is controlled through a synchronous dq-axis rotating frame. There are multiple choices when applying directional rotating frames. Traditionally, in induction machines, rotor-flux orientation and magnetizing-flux were utilized in vector control [5]. However, the orientation of the d-axis along the stator flux vector position is one of the most common implementation methods in recent DFIG. This method is called stator-flux orientation (SFO) vector control. Another common orientation method is the stator-voltage orientation (SVO) [6]. In the SFO method, it is possible to independently control electrical torque and rotor excitation current. Thus, controlling the active and reactive power.

The rotor-side converter's main purpose is to supply the excitation for the DFIG rotor. This is solely achieved by controlling the rotor currents such that the rotor flux

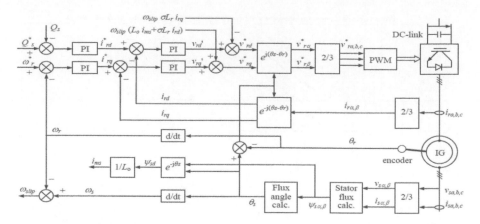

FIGURE 10.4 Control Structure of the rotor-side converter [7].

is optimally oriented with respect to the stator flux. Therefore, the integrated pulse-width modulation (PWM) converter is enabled to control the torque and, hence, the speed of the DFIG.

In Figure 10.4, a schematic block diagram depicts the control structure of the rotor-side converter. The same is modeled in RTDS.

The equations governing the relationship of rotor voltage (v_d and v_q) and flux (ϕ_d and ϕ_q) components in the dq-axis and under the SFO are shown as follows:

$$v_{rd} = R_r i_{rd} + \sigma L_r \frac{di_{rd}}{dt} - \omega_{slip}\sigma L_r i_{rq} \tag{10.2}$$

$$v_{rq} = R_r i_{rq} + \sigma L_r \frac{di_{rq}}{dt} + \omega_{slip}\left(L_o i_{ms} + \sigma L_r i_{rd}\right) \tag{10.3}$$

$$\phi_{rd} = L_o i_{ms} + \sigma L_r i_{rd} \tag{10.4}$$

$$\phi_{rq} = \sigma L_r i_{rq} \tag{10.5}$$

where $\sigma = 1 - \frac{L_m^2}{L_s L_r}$ and $L_o = \frac{L_m^2}{L_s}$, and L_r, L_s, and L_m are the rotor and stator self-inductances, and magnetizing inductance, respectively.

As seen in Figure 10.4, the rotor-side converter consists of two series of proportional-integral (PI) controllers in parallel forming two control loops. In the reference q-axis rotor current i_{rq}^*, the outer loop can be a speed-control loop or a torque-control loop. However, for the torque-control, it is usually difficult to precisely measure the machine toque. Hence, when torque is used, it becomes an open loop. The outer control loop is indirectly connected to the output of active power (P). Moreover, additional PI-controllers are used to produce the d-axis rotor current reference component i_{rd}^* to regulate the required reactive power Q.

The controller requires the measurements of the stator voltage and current as well as the rotor current. Finally, the rotor excitation current control is achieved by regulating the rotor voltage output induced by the VSI. The i_{rd} and i_{rq} error signals are fed into the respective PI controllers to generate the required v_{rd} and v_{rq}, respectively.

$$v'_{rd} = R_r i_{rd} + \sigma L_r \frac{di_{rd}}{dt} \tag{10.6}$$

$$v'_{rq} = R_r i_{rq} + \sigma L_r \frac{di_{rq}}{dt} \tag{10.7}$$

10.2.2 DFIG GRID-SIDE CONTROL

The grid-side converter (GSC), in DFIG, controls the flow of active and reactive power to the grid. Its main objective is to maintain a constant voltage at the direct current (DC)-link irrespective of the rotor power value and direction.

To allow an independent active and reactive power control between the grid and the GSC, the rotating reference frame is oriented along the stator voltage vector. The PWM converter is current controlled, with the d-axis current used to regulate the DC-link voltage and the q-axis current component to regulate the reactive power. A schematic of the GSC control structure is displayed in Figure 10.5.

To achieve a decoupled control of active and reactive power, compensation terms are inserted in the v_d and v_q equations as follows:

$$v_{cd} = R i_{cd} + L_{choke} \frac{di_{cd}}{dt} - \omega_e L_{choke} i_{cq} + v_{cd1} \tag{10.8}$$

$$v_{cq} = R i_{cq} + L_{choke} \frac{di_{cq}}{dt} + \omega_e L_{choke} i_{cq} + v_{cq1} \tag{10.9}$$

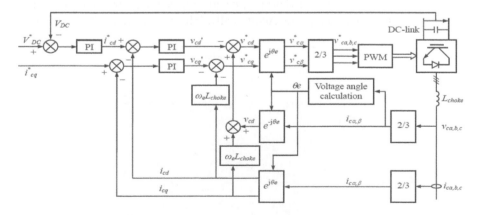

FIGURE 10.5 Control structure of the grid-side converter [7].

Rotor-Side Converter Hybrid_Plant

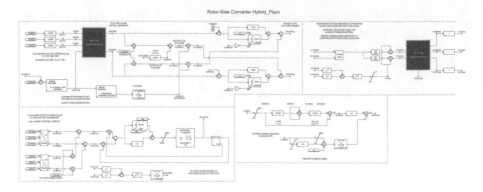

FIGURE 10.6   Schematic of rotor-side converter control in RTDS.

where θ_e is the grid voltage angular position and

$$v'_{cd} = R\, i_{cd} + L_{choke}\, \frac{di_{cd}}{dt} \tag{10.10}$$

$$v'_{cq} = R\, i_{cq} + L_{choke}\, \frac{di_{cq}}{dt} \tag{10.11}$$

The control models of the rotor-side converter and GSC modeled in RTDS are displayed in Figures 10.6 and 10.7, respectively.

10.3   DEVELOPING HARDWARE-IN-THE-LOOP (HIL) STRUCTURE

HIL testing provides a useful simulation environment to develop, analyze, and test designed control schemes for complex and detailed systems.

In HIL, a computer along with a real-time simulator can be utilized to run repetitive procedures to test and modify the control model in the study. A plant model running in a real-time simulator (e.g., RTDS) is directly connected to a hardware or a control system. Hence, it is possible to test and validate the control scheme prior to prototyping and production phase. Additionally, HIL is a superb method for testing

Grid-Side Converter Hybrid_Plant

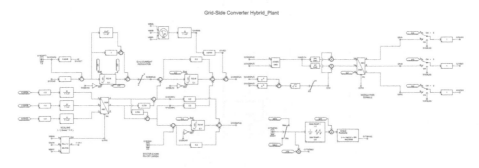

FIGURE 10.7   Schematic of grid-side converter control in RTDS.

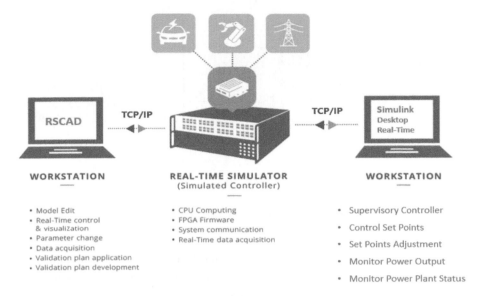

FIGURE 10.8   Hardware-in-the-loop components interaction.

new modifications and upgrades before implementing them into the real-time system by detecting possible defects, bugs, inaccurate algorithms, and unanticipated behaviors [8] and [9].

In HIL simulation, the interaction between the simulation system and the tested controller is being mimicked so that the inputs to the controller appear to be obtained from a real system and the control signals and commands from the controller cause corresponding actions in the model [10] and [11].

As presented in Figure 10.8, an HIL structure is developed, in which, the hybrid power plant is modeled in RTDS and the central controller is modeled in Simulink Desktop Real-Time (SLDRT). The central controller performs computations in real-time and sends control commands to the multiple various inverters that are part of the hybrid plant model. The simulation itself is monitored by RSCAD RunTime (i.e., a package part of RSCAD). RSCAD RunTime runs on a separate workstation to provide better performance and to not load the processor with other tasks. Along with monitoring, it controls the simulation and extracts the results and analysis data in very high resolution reaching one sample every 50 µs per signal as well as storing them in different formats for transient studies. As for steady-state analysis, data acquired by Simulink are sufficient with a sampling rate of 1 kHz. The whole system is synchronized with the hybrid power plant model running inside RTDS.

10.4   DEVELOPING CENTRAL CONTROLLER IN SIMULINK

As previously explained in Chapter 1, the local controllers attached directly to the inverters are considered as primary controllers, whereas the central controller modeled in Simulink is considered as a secondary controller (see Figure 10.9). The primary goal of the proposed central controller is to track a reference for active and

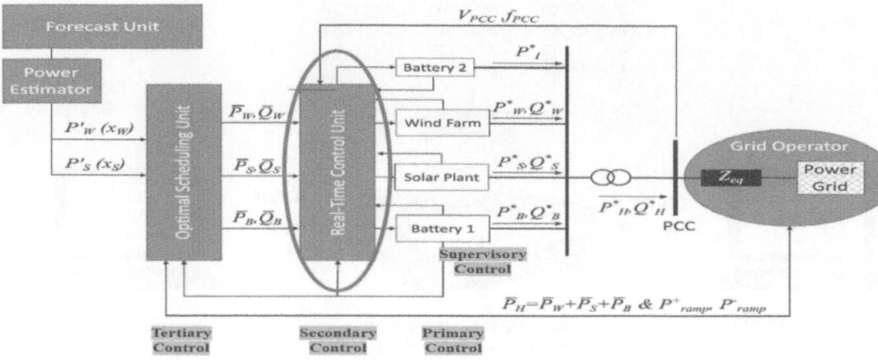

FIGURE 10.9 Hierarchical control in a hybrid power plant.

reactive power at the point of common coupling (PCC) for a specific time interval set by the electricity spot market rules. The reference power at PCC is aggregated by a separate reference power at every local controller. The reference could be set either by the energy management system (EMS) tertiary controller or in some special cases by the secondary controller itself, depending on the status of different components in the hybrid power plant. In such a special case, the tertiary controller is notified about any reference adjustments as the secondary controller has a faster response.

The central controller also monitors the battery state of charge (SoC) and controls the charging/discharging operations. Additionally, the central controller assigns one of the local controllers as a supervisory controller. The objective of the supervisory controller is to correct any residual power mismatch at the PCC; hence, its reference input is the calculated error signal of the PCC, therefore, achieving a more stable operation as primary controllers have the fastest response in the system.

The structure of the central controller is illustrated in Figure 10.10. The main components of the central controller are the receiver and transmitter, data processing unit, battery dispatching unit, and the main dispatching unit. As displayed in Figure 10.10, the central controller has two main communication links, one with the simulated hybrid plant in RTDS via TCP/IP and the second link with the EMS System, which in turn communicates with the grid operator.

The interactions between different control levels in the hybrid plant are demonstrated by Figure 10.11. The committed dispatchable power is processed by the tertiary controller (i.e., EMS) and power reference is adjusted by the scheduling controller. Next, the battery dispatching unit compensates the reference dispatch power to produce the reference power output of individual resources in the hybrid plant and the expected total power at the PCC. The assigned supervisory controller then switches its operating mode to follow the PCC output power error signal as its reference. It is worth mentioning that primary controllers operate in real time with a response time of around 100 μs. The secondary controller is modeled to have a variable response time between 10 ms up to 1,000 ms. However, it can be reprogrammed to have an even slower response time reaching 10 s. The tertiary controller, on the other hand, has the slowest response in normal operation, in the range of a few

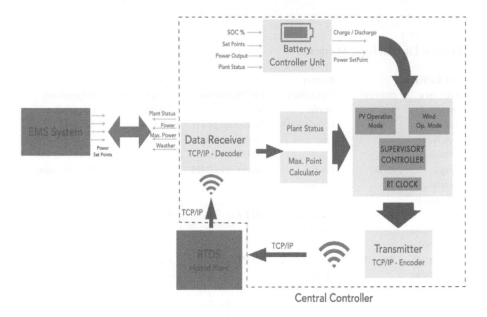

FIGURE 10.10   Central Controller Structure.

minutes. Having said that, any abnormal interrupts in the system are communicated with different control levels instantaneously through status data points.

10.4.1   BATTERY DISPATCHING ALGORITHMS

The battery dispatching unit uses input data from the SoC, dispatching information from the EMS, status data of different renewable sources, and hybrid plant status and current power output from all renewable sources to determine the best battery operation mode along with required power output set point. Table 10.5 illustrates the available battery dispatching modes.

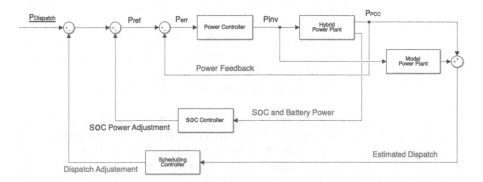

FIGURE 10.11   Hybrid power plant hierarchical control loops.

TABLE 10.5
Battery Dispatching Modes

EMS Battery Reference Power	Battery SoC	Battery Operation Mode	PV Operation Mode	DFIG Operation Mode	Supervisory Controller
Charge/ Discharge (Normal)	20%–80%	Follow EMS ref	Maximum power point tracking (MPPT)	Max. power	Battery
Charge/ Discharge (Abnormal)	<20%	Priority to charge	Supervisory/MPPT	Max. power	One PV string
Charge/ Discharge (Abnormal)	>80%	Priority to discharge	MPPT	Max. power	Battery
Compensate Lost Power (Emergency)	10%–100%	Discharge up to max. rate	MPPT	Max. power	Battery (up to max. limit)
Curtailment (Emergency)	0%–95%	Charge up to max. rate	MPPT/curtailment	Reduced power	Battery (up to max. limit) then use curtailment

To explain Table 10.5, the normal operation is identified when the SoC is between 20% and 80% and there is no abnormal status in the plant. However, those SoC limits can be varied by the optimal dispatching unit.

In normal operation, the battery dispatch is calculated by the optimal scheduling unit (EMS) and executed by the central controller. When the battery is set to discharge, other resources are set to operate on maximum output power to ensure the battery discharge rate is minimized. Similarly, when the battery is set to charge, other resources are set to maximum power mode to utilize all available power up to maximum charge rate then, if required, PV output is reduced below maximum point to maintain a constant dispatched power at PCC. Moreover, the battery set point calculated by the EMS controller is not strictly followed by the central controller; instead, battery output power would be floating around the set point to compensate for any forecasting error and deal with any fluctuations caused by other renewable resources (i.e., the main objective is to keep hybrid plant power output constant at contracted dispatched power). Hence, the battery local controller is assigned as a supervisory controller monitoring the plant total output power at PCC.

In abnormal situations, when SoC is more than 80% or less than 20%, the central controller is allowed to discard the EMS battery set point and make charging or discharging a priority, as per Table 10.5. In such an operation, the EMS controller is notified by a status signal to adjust its calculations. In this state, the supervisory controller may be switched to one of the PV strings while leaving other strings to operate in MPPT mode.

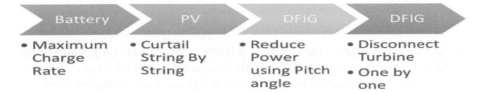

Battery	PV	DFIG	DFIG
• Maximum Charge Rate	• Curtail String By String	• Reduce Power using Pitch angle	• Disconnect Turbine • One by one

FIGURE 10.12   Curtailment process.

In emergency situations such as losing one of the available resources, the battery SoC limits are relaxed and the battery is forced to compensate for lost/excess power by using maximum discharge/charge rate until the EMS unit readjusts its calculated set points.

In case a curtailment is required, PV strings are curtailed first due to their uncomplicated synchronization process and quick operation. However, DFIGs have high inertia and their curtailment process is relatively long; hence, their power output is reduced by controlling the pitch angle. Then, if further reduction is required, DFIGs are disconnected. The curtailment process is illustrated in Figure 10.12.

10.4.2 SUPERVISORY CONTROL

A supervisory controller is one of the primary local controllers operating in a special mode to compensate for the total hybrid plant power output. Its input reference is the power error signal at PCC. Most of the time, the battery controller is assigned as a supervisory controller. The equations below describe the input reference signal of the supervisory controller.

$$P_{pcc\_ref} = P_{res\_tot} + P_{super} \tag{10.12}$$

where P_{pcc_ref} is the reference of the plant total dispatched power and P_{res_tot} is the actual dispatched power from renewable resources measured at the PCC.

When the battery is in supervisory mode

$$P_{res\_tot} = P_{pv\_tot} + P_{DFIG\_tot} \tag{10.13}$$

$$P_{ESS} = P_{super} \tag{10.14}$$

hence,

$$P_{ESS} = P_{error} = P_{pcc\_ref} - P_{res\_tot} \tag{10.15}$$

10.5   TESTING AND SIMULATION RESULTS

This section describes the performed tests and simulations along with their acquired results. The results are then analyzed and presented.

10.5.1 PERFORMANCE OF INDIVIDUAL PRIMARY CONTROLLERS AND OPERATION MODES

In the hybrid power plant, the inverters' primary controllers are programmed with multiple operation modes to facilitate the execution of the hierarchical control scheme. The main operation modes for real power are tracking maximum power point or maintaining a constant power output. Similarly, for reactive power, the main operation modes are either regulating the voltage of the alternating current (AC) bus or maintaining a constant output value set by the reference input. Additionally, there is the supervisory control mode where the output of the energy source is continuously varying depending on the state of PCC power. The different operation modes for different energy sources are listed in Table 10.6. From the table, it is obvious that DFIG has the least number of operation modes due to its inertia, slower response, and other inherited mechanical characteristics.

All the mentioned modes need to be tested individually and collectively before they can be integrated with the secondary controller in the hierarchical control scheme.

The PV controllers are tested in a simulation to check the response of both the maximum power point tracking (MPPT) mode and the constant power mode separately. Figure 10.13 illustrates the response of the MPPT mode while varying irradiance level; whereas Figure 10.14 displays the operation of the PV controllers when switched to constant output power mode. The supervisory mode is tested thoroughly in Section 10.5.3. In the MPPT mode, the PV primary controller displayed a fast response following the maximum power point in reaction to fluctuating irradiance level. This fast response offered by inverter-based energy sources qualifies PV strings to act as a supervisory controller when the battery system is unavailable.

The PV response depicted in Figure 10.14 illustrates that the PV controllers are capable of producing a constant output power despite the changing sun irradiance. In this test, PV1 and PV3 are initially set to different power reference 0.6 and 0.5 MW, respectively. However, PV2 is operated in MPPT to compare it with other PV strings.

TABLE 10.6
Different Energy Sources Operation Modes

Operation Mode	Battery Energy System	PV Strings	DFIG
Max. active power (P)	Yes	Yes	Yes
AC voltage regulation	Yes	Yes	No
Constant active power (P)	Yes	Yes	Yes (limited)
Constant reactive power (Q)	Yes	Yes	Yes (limited)
Variable active power (P)	Yes	Yes	No
Variable reactive power (Q)	Yes	Yes	No
Supervisory control	Yes	Yes	No
Emergency mode	Yes	No	No

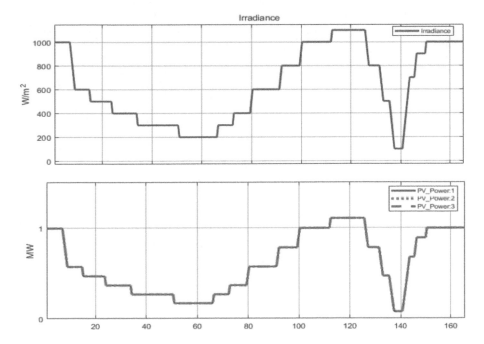

FIGURE 10.13 PV MPPT mode test.

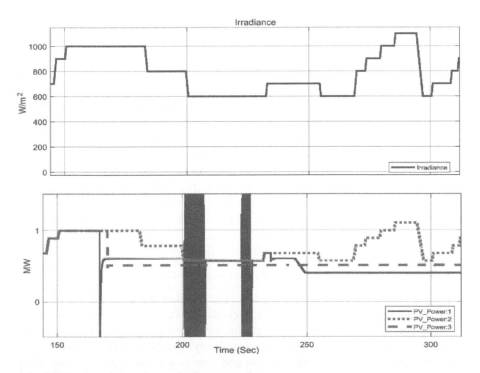

FIGURE 10.14 PV Fixed power reference mode test.

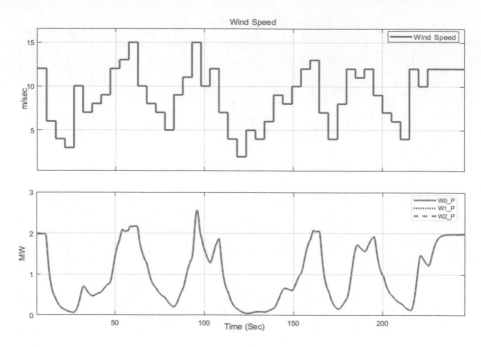

FIGURE 10.15 DFIG maximum power mode test.

At simulation time 200, the irradiance level drops below the level where PV1 can produce its reference power. Consequently, PV1 crosses into an unstable zone and an oscillating behavior is observed. This phenomenon is discussed in detail in this report. At the end of the test, the PV1 reference was lowered to 0.4 MW. It can be concluded from this test that, when PV controllers are operated in supervisory and constant output power modes, the reference values must be limited by the simultaneous maximum available power.

The independent tests of DFIG controllers are presented in Figures 10.15 and 10.16. Figure 10.15 displays the behavior of DFIG in maximum power mode while varying wind speed. On the other hand, Figure 10.16 demonstrates the response of the DFIG when the reference power value was decreased below the maximum available power. It can be observed that DFIGs have a slow response; therefore, they are not suitable to be operated in a supervisory capacity, which requires a quick reaction to sudden and continuous fluctuations.

10.5.2 Communication Link with Secondary Controller

As previously illustrated, a communication link is set up between the hybrid power plant model running in RTDS and the secondary controller modeled in Simulink. For the purpose of this project, the communication application requirements are basic, e.g., the ability to exchange data seamlessly with no specific communication protocol type requirement, but a simple and easy to use communication interface would be desirable.

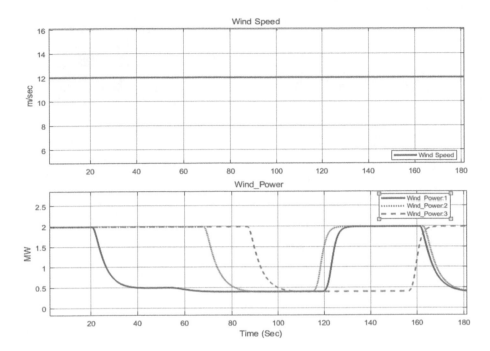

FIGURE 10.16   DFIG constant power mode test.

The GTNET-SKT protocol can address the communication needs from the afore-mentioned case. It is based on the familiar TCP socket communication program model. Hence, any external application can connect with RTDS using the protocol as long as it supports the socket programming model displayed in Figure 10.17. In addition to its simplicity, the TCP socket protocol offers a sufficient communication speed.

To establish the communication link, in RSCAD, the GTNET-SKT model representing the GTNET-SKT communication interface card is utilized (see Figure 10.18). In this project, the GTNET-SKT takes the role of the socket server, whereas the Simulink model plays the role of the socket client. The communication initiates from the client side first. Once the socket is established, the GTNET-SKT starts to send the data points to the Simulink client at a given update frequency. The model controlling the transmit frequency is displayed in Figure 10.19.

The transmission frequency can change while the simulation case runs, using a slider in the simulation case. The Simulink model receives the incoming data packets from the GTNET-SKT and extracts those data points. Then, the Simulink model feeds the incoming data into the secondary controller model to calculate the new reference set points as well as for deciding on the operational modes for every energy source. Once new commands from the secondary controller are ready, the client in Simulink sends back those new commands to the server, which is the GTNET-SKT.

To test the performance of the developed communication link in RTDS and Simulink, a set of 20 data points is exchanged back and forth over a period of time while varying the exchange frequency. The communication average delay is measured by a model in RTDS and displayed while the simulation is running.

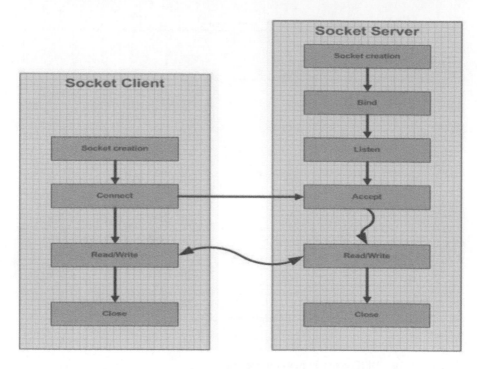

FIGURE 10.17 TCP socket programming model.

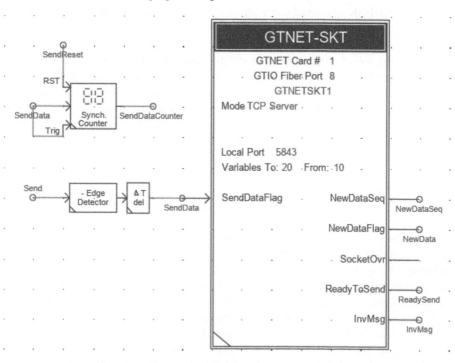

FIGURE 10.18 GTNET-SKT model in RSCAD.

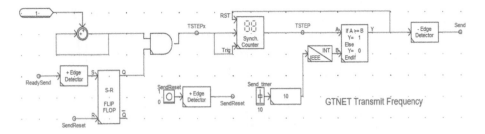

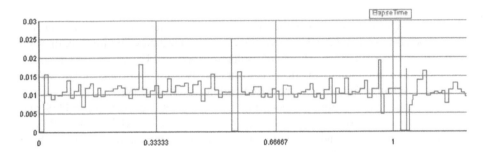

FIGURE 10.19 GTNET transmit frequency model.

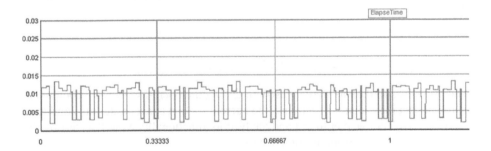

FIGURE 10.20 Communication test 1.

FIGURE 10.21 Communication test 2, optimized Simulink code.

The first results are demonstrated in Figure 10.20 where the vertical axis represents the elapsed time in seconds. As displayed in the first results, the average communication delay was around 15 ms and some data packets took a lot longer.

As a result, the code manipulating the incoming data in Simulink was optimized by deleting unnecessary data conversions and data masking (i.e., RTDS sends data as a "single" variable where every value is masked in 4 bytes).

After optimizing the code in Simulink, the second test results are displayed in Figure 10.21. The elapsed time was improved; it was reaching a few milliseconds but most of the time it was above 10 ms.

To further enhance the communication link, the Simulink model was enabled to take control of the computer processor to execute in real time in the external mode. In external mode, Simulink Coder™ is used to generate a code combining the

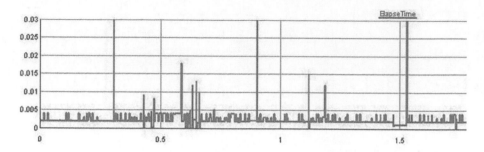

FIGURE 10.22 Communication test 3 with Simulink real-time external mode.

simulation algorithms as well as the I/O drivers. The generated code is run entirely in the operating system kernel and exchanges parameter data with Simulink through a shared memory interface. The final results are shown in Figure 10.22. The graph illustrates that communication delay is around only 2 ms and some spikes appear when the exchange frequency is increased beyond 2 milliseconds.

To conclude, the communication link is established between the RTDS and Simulink model utilizing TCP socket protocol and the GTNET card in RTDS. When modeling the secondary controller in Simulink using the Real-Time Desktop toolbox, the best communication performance is achieved when the model is running in the external mode where Simulink takes control of the computer processor. Additionally, a real-time operation is assured with the synchronization realized by RTDS.

10.5.3   TESTING THE COMPLETE CONTROL STRUCTURE PERFORMANCE WITH REAL-TIME DATA SIMULATIONS

The previous simulations were focused on testing the operation of every component of the project individually. Therefore, the current section concentrates on simulating the whole project together to test and analyze its functionality as well as testing the interactions between different components.

10.5.3.1   The Case Study

Harweel wind farm is the first large-scale wind farm power station in Oman. It is located in Dhofar Governorate, south of Oman, and connected with Dhofar Grid as well as the Petroleum Development of Oman (PDO) Grid. This 50-MW wind farm consists of 13 × 3.8-MW wind turbines and was commissioned at the beginning of 2020. With the beginning of operation, it opened new opportunities for renewables integration into Oman's power system. It is projected that the Harweel wind farm will generate approximately 167 GWh clean energy per year; this is equivalent to a 38% capacity factor [12]. However, because of the intermittent behavior of wind speed, the power generated in the Harweel wind farm fluctuates.

In [12], 1-year 10-minute resolution historic wind speed and direction measurements were taken at the Harweel wind farm site, and the wind farm layout and turbine properties were used to calculate the power output of the wind farm.

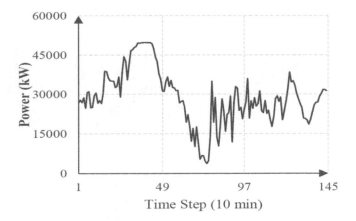

FIGURE 10.23 Harweel wind farm output for a typical summer day [12].

Figure 10.23 illustrates the calculated power output of Harweel wind farm for a typical summer day.

As seen in Figure 10.23, the power generated from the wind farm continuously fluctuates throughout the day. This is the same for the rest of the days in the year. The fluctuations can be eliminated by using the combination of wind and solar energy integrated with the advantages of the battery storage system.

The presented EMS for the dispatchable hybrid renewable generation power plant may be utilized to balance out the fluctuations of power output and deliver a dispatchable power from such a hybrid wind-solar power plant.

A case study was carried out by [12] to demonstrate the potential of the Harweel wind farm for delivering dispatchable power. One-year 10-minute resolution historic data for wind speed and direction as well as solar irradiance [3] at the Harweel wind farm site were used. The output of the power estimator was calculated by [12] using the forecast data. The proposed optimal scheduling unit in [12] was used to find set points for the wind farm, solar PV system, and BESS such that dispatchable power can be achieved. The trading period length was assumed to be 30 minutes.

The results of the case study are displayed in Figure 10.24. These results were subsequently employed by this project to test the functionality of the proposed hierarchical control system scheme. However, because the capacity of the modeled hybrid power plant in RTDS is different from the one presented in [12], the data were first normalized, per capacity, and then readjusted to fit the modeled power plant and run the simulation in RTDS. The forecast data, along with calculated set points of different renewable resources in the hybrid power plant, are presented in Figure 10.25.

10.5.3.2 Scenario 1

In the first simulation, the main objective was focused on testing the operation of the control system during early morning while PV power output is starting to grow as the sun irradiance increases gradually. Sub-objectives were included in the simulation to examine the response of the control system in different situations when different scenarios are applied.

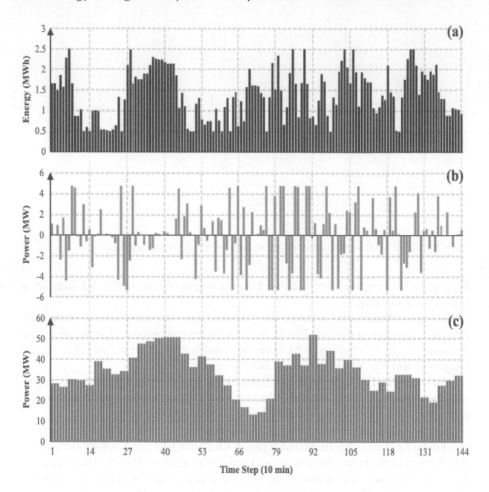

FIGURE 10.24 Output of dispatchable hybrid power plant for a typical summer day. (a) BESS stored energy. (b) BESS power output. (c) Dispatchable power [12].

These sub-objectives include:

- Testing the supervisory mode of battery system controller
- Testing the response of faults in one and two PV strings
- Testing the supervisory mode of PV controller

The results shown in Figure 10.26 are part of the full simulation that is presented separately at the end of this chapter. The simulation is fast-forwarded 25 times for display purposes; hence, every 25 s are represented by 1 s in the graphs. Figure 10.26 illustrates the operation of the battery controller supervisory mode as well as the response of the control system when one or more of the renewable resources are lost.

Just before simulation time 80, PV0 is lost and followed by PV1 after a few minutes. After the first failure, the battery controller compensates the lost power by discharging the battery. However, when the second PV string is lost, the available

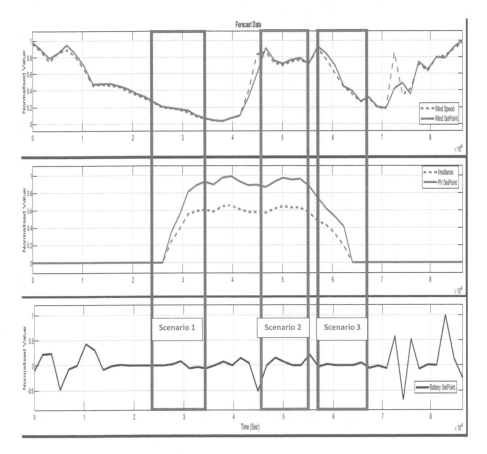

FIGURE 10.25 Normalized values of forecast weather parameters and set points of the renewable resources in the hybrid power plant.

resources in the power plant cannot further compensate the lost power as maximum output is reached. Hence, a signal is sent to notify the secondary controller about the situation. The secondary controller, then, contacts the tertiary controller (i.e., EMS optimal scheduling unit) to readjust the total power plant reference point due to the emergency situation. The reference point is readjusted as displayed in Figure 10.26. When the lost resources are restored, the power plant is brought back to normal operation.

The second part of the simulation is displayed in Figure 7.27, which highlights the operation when the supervisory mode is switched to PV. At simulation time 220, the battery is forced to stop to simulate the transition of the supervisory mode from battery controller to PV1 controller. When the PV1 controller is in supervisory mode, it reduces the output of PV1 to match the required power output of the hybrid plant at the PCC point. However, in case of emergency, if another resource is lost, the battery is brought back to service to compensate for lost power, as demonstrated in Figure 10.27, and according to Tables 10.5 and 10.6. In such a

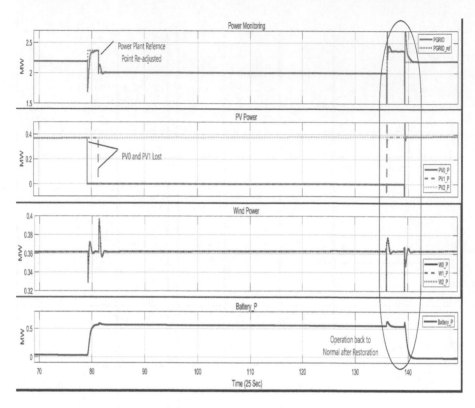

FIGURE 10.26 Simulation 1, part 1, battery supervisory mode and PV failure response.

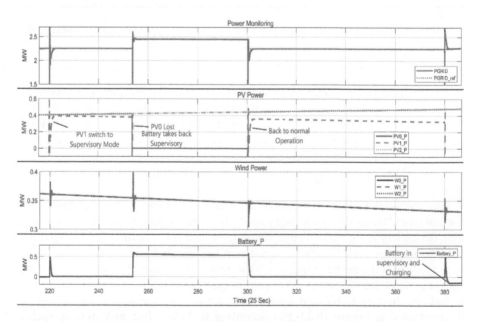

FIGURE 10.27 Simulation 1, part 3, PV in supervisory mode.

situation, PV1 is operated at maximum power mode to reduce the discharging stress on the battery.

Once the lost resource is restored, the battery is switched off, the power plant operation goes back to normal, and the PV1 controller returns to supervisory mode. This continues following the power plant reference points until the battery controller is allowed to start again.

The full duration of simulation 1, lasting almost 3 hours, is displayed on Figure 10.28.

10.5.3.3   Scenario 2

The second simulation concentrates on diagnosing the behavior of the PV controller in the unstable zone.

In normal circumstances, PV controllers are operated in the MPPT mode. In MPPT mode, a PV controller is prevented from entering the unstable zone by the MPPT algorithms. However, when a PV controller is operated in other modes, it can easily cross inside the unstable zone and gets trapped inside it unless a mechanism is developed to prevent such behavior. Figure 10.29 illustrates the results when a single PV controller enters the unstable zone. The output of the power plants swings heavily

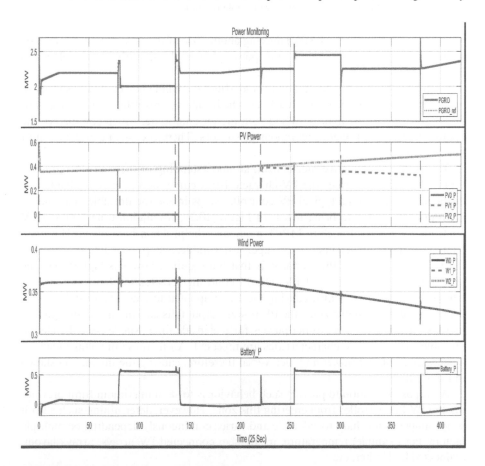

FIGURE 10.28   Simulation 1 full duration.

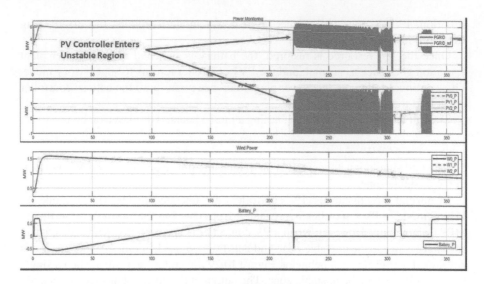

FIGURE 10.29 Simulation 2, controller unstable operation.

with a huge amplitude (3–4 MW) making about 70% of the total output power. The operation of wind turbines in the same power plant is also disturbed, as displayed in Figure 10.29.

Such behavior is very critical and could affect the whole power grid, resulting in catastrophic impacts and even a blackout. The harmonics resulting from such conditions propagate in the main grid and gain higher amplitude, which can disturb the operation of conventional synchronous generators. Therefore, such conditions must not be allowed to occur.

To develop a preventive mechanism, an understanding of the cause is required. Figure 10.30 demonstrates the PV characteristics curve along the controllable and unstable zones. When a typical PV controller steps inside the unstable zone then it is almost impossible to bring it out. In Figure 10.29, simulation time 300, the PV controller is switched back to MPPT mode and the problem is resolved. However, later, when it is switched back to the supervisory mode, it steps back into the unstable zone. The main reason this happens is that the required power is higher than the available maximum power of PV. The PI controller creates a reference i_d and i_q, which are outside the limits, causing the V_d component to enter the unstable zone. When inside the unstable zone, the PV power output falls causing the error signal to increase, hence, the PI controller pushes further inside the unstable zone. Inside the unstable zone, the P–V characteristics are reversed. An increase in voltage leads to a decrease in output power and vice versa; therefore, causing the fluctuation shown in Figure 10.29.

A proposed solution to prevent such behavior is to set a maximum limit preventing the PV PI controller from entering this zone. However, determining such a limit is not simple as it is hard to calculate and varies continuously depending on multiple factors. For example, temperature, irradiance, connected PV arrays, arrays layout, connected LCL filter, etc.

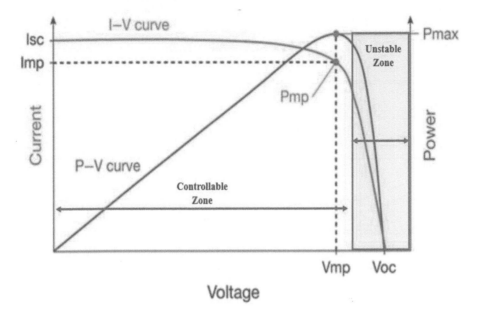

FIGURE 10.30   PV unstable zone.

A proposed solution was to derive the limit from one of the other identical online PV strings because they have the same characteristics and they are affected by the same conditions. The result of this simulation is demonstrated in simulation 3.

10.5.3.4   Scenario 3

Simulation 3 represents the results of the modified PV controller to avoid operating in the unstable zone. Additionally, it aims to test the new improved PI controller in supervisory mode.

The simulation results are demonstrated in Figure 10.31. It is clear that the supervisory mode is prevented from entering the unstable zone. However, when higher reference power is requested, the battery is brought back to service. In rare cases, where the reference requested power is higher than the total maximum power, all resources are operated at the maximum point and a signal is sent back to the tertiary controller notifying it to readjust its set points. It is worth noting that this is a hypothetical scenario that should not occur if all EMS components are working properly along with the forecasting engine. However, such a condition has to be planned for and tested in case of emergencies such as a failure in one of the hybrid power plant components.

10.5.4   Simulating A Communication Loss

In this scenario, a communication loss is simulated to analyze the operation of the hybrid power plant in such a critical condition. The secondary and primary controllers are both programmed to follow the latest received power set points. Consequently,

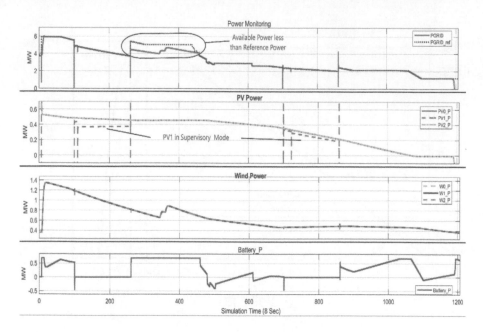

FIGURE 10.31 Simulation 3, improved PV PI controller.

when a communication is lost, the reference power output will be unchanged during this period. However, weather conditions may vary during this period causing the actual available power output to be less than the last updated reference power, as demonstrated in Figure 10.32. Later, when communication is restored, the reference power would reflect the actual available power. As depicted in Figure 10.32, after

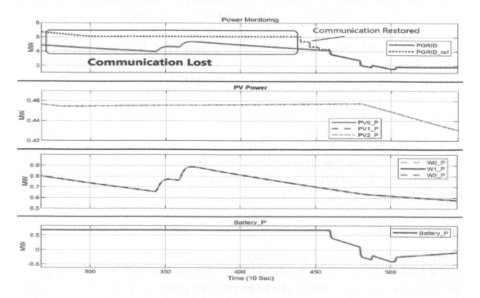

FIGURE 10.32 Simulation of lost communication.

the communication is restored, the actual generated power started following the new updated reference power set points.

10.6   ONLINE INTEGRATION BETWEEN EMS AND RTDS

This section provides details and concise descriptions on how to integrate the developed EMS tools in Chapter 9 with the RTDS. This shows how the developed power dispatching system based on the optimization method can be used for online update of the control set points in secondary controllers used in real-world power systems in control centers or virtually implemented in RTDS, as described in previous sections in this chapter. Therefore, this section develops and implements an SQL database with a communication link with RSCAD used in RTDS facilities.

10.6.1   BACKGROUND ON THE LINK BETWEEN SQL AND RTDS AND OBJECTIVES

The energy produced from wind turbines and PV systems is not reliable by itself due to its intermittent nature. As a result, wind and PV resources are nondispatchable. Integrating wind and PV economically into a conventional power grid necessitates the elimination of the complications caused by the intermittent behavior of these resources.

As seen in the previous sections and chapters, our research team at SQU has executed a project to develop an EMS for dispatchable renewable power generation. The project is divided into four main components: forecast unit, power estimator, optimal scheduling unit, and real-time control unit. Likewise, comprehensive work has been done on designing, developing, and simulating the real-time control unit in our project, as fully described in the previous sections. Briefly, our team has developed a hierarchical control system (central control unit) and its working algorithms to operate the hybrid plant and execute the set points of a higher EMS optimal scheduling unit. The designed hierarchical control scheme was simulated in an RTDS system with different scenarios. However, the set points for the control system simulations were generated in advance and loaded at the beginning of each simulation and stored in the controller memory where they can't be varied during the simulations.

To make the project realistic for real-world implementations and to mimic the control process in control rooms in real-world power systems, an objective was set to build a database structure to store the instructions and set points of the optimal scheduling unit and accessed in real time to update, add, and modify these set points. The previously developed control system is connected to the database to fetch the up-to-date set points and execute them in the simulations in real time. Therefore, the main objectives of this section include the following:

1. Develop a dynamic SQL database to store the instruction and set points that are required for the execution of the control system and the hybrid power plant simulations
2. Develop a communication link between the SQL database and the control unit to facilitate data transfer in real time
3. Develop a dynamic SQL query that can automatically fetch the required set points from the SQL database in real time whenever it is called by the control unit algorithms during a simulation

4. Modify the algorithm of the hierarchical control unit to set a periodic trigger to update the hybrid power plant set points and execute the currently loaded set points depending on the simulations run time

10.6.2 Developing SQL Database

An SQL server has been configured in the local workstation that is running MATLAB®. Moreover, an SQL database has been created with the name "RTDS" using the Microsoft SQL Server Management Studio. It can be migrated later to an independent remote server or the Azure Cloud service. The RTDS database currently consists of four tables figure 10.33. Table "Res_Forecast" stores the weather forecast along with the forecasted power outputs of the available renewable energy sources according to the forecasted weather. The other three tables store the set points of each renewable source, which is calculated by the optimization problem in the optimal scheduling unit. The set points are stored as normalized values to accommodate any resource capacity in the simulations. Moreover, a separate column has been added to every table that refers to MATLAB simulation time figure 10.34. This column will facilitate the execution of MATLAB queries according to the simulated scenarios.

10.6.3 ODBC Link

Database vendors, such as Microsoft, implement their database systems using technologies that vary depending on customer needs, market demands, and other factors. Software applications written in popular programming languages, such as C, C++, and Java®, need a way to communicate with these databases. Open database connectivity (ODBC) and Java database connectivity (JDBC) are standards for drivers that enable programmers to write database-agnostic software applications. ODBC and JDBC provide a set of rules recommended for efficient communication with a database.

ODBC is a standard Microsoft Windows® interface that enables communication between database management systems and applications typically written in C or C++. MATLAB Database Toolbox™ has a C++ library that connects natively to an

	TABLE_CATALOG	TABLE_SCHEMA	TABLE_NAME	TABLE_TYPE
1	Rtds	dbo	sysdiagrams	BASE TABLE
2	Rtds	dbo	Battery	BASE TABLE
3	Rtds	dbo	PV	BASE TABLE
4	Rtds	dbo	Res_Forecast	BASE TABLE
5	Rtds	dbo	Wind	BASE TABLE

FIGURE 10.33   RTDS database with tables.

```
/****** Script for SelectTopNRows command from SSMS ******/
SELECT TOP (100) [DT]
      ,[Wind_SP]
      ,[Wind_SP_Norm]
      ,[MatlabTime]
  FROM [Rtds].[dbo].[Wind]
```

121 %

⊞ Results Messages

	DT	Wind_SP	Wind_SP_Norm	MatlabTime
1	2006-02-14 00:00:00.000	48.02803015	0.960560603	0
2	2006-02-14 00:01:00.000	48.1393936	0.962787872	0.0166666666666667
3	2006-02-14 00:09:00.000	48.1393936	0.962787872	0.15
4	2006-02-14 00:10:00.000	46.9488635	0.93897727	0.166666666666667
5	2006-02-14 00:11:00.000	45.7583334	0.915166668	0.183333333333333
6	2006-02-14 00:19:00.000	45.7583334	0.915166668	0.316666666666667
7	2006-02-14 00:20:00.000	44.69270835	0.893854167	0.333333333333333
8	2006-02-14 00:21:00.000	43.6270833	0.872541666	0.35
9	2006-02-14 00:29:00.000	43.6270833	0.872541666	0.483333333333333
10	2006-02-14 00:30:00.000	43.4360336	0.868720672	0.5
11	2006-02-14 00:31:00.000	43.2449839	0.864899678	0.516666666666667
12	2006-02-14 00:39:00.000	43.2449839	0.864899678	0.65
13	2006-02-14 00:40:00.000	41.91520025	0.838304005	0.666666666666667

FIGURE 10.34   Sample of wind table entries in RTDS database.

ODBC driver. The advantages of ODBC are the fastest performance for data imports and exports and memory-intensive data imports and exports. The created ODBC interface is configured to directly link to the RTDS database in the local SQL server for a smoother and faster performance figure 10.35.

10.6.4   DEVELOPING DYNAMIC SQL QUERY

Data import or export to the SQL database is achieved through executing SQL queries, which is basically a language used by databases. This language handles the information using tables and shows a language to query these tables and other related objects (views, functions, procedures, etc.).

Single query has been created and tested to fetch all the required data by MATLAB from all the tables in the RTDS database; hence, minimizing the number of executed query instructions during the real-time simulations to avoid affecting the performance while waiting for the SQL server to handle the query. The query utilizes inner joints between different tables. It also uses the "matlabtime" column to fetch only the current required data according to the simulation time. An example of the query is displayed in Figure 10.36.

The query is set dynamically by reading the simulation time as an input and automatically setting the range of the required data by changing the range of "MatlabTime" field. In this case it is 10 to 15 minutes, as discussed in next section.

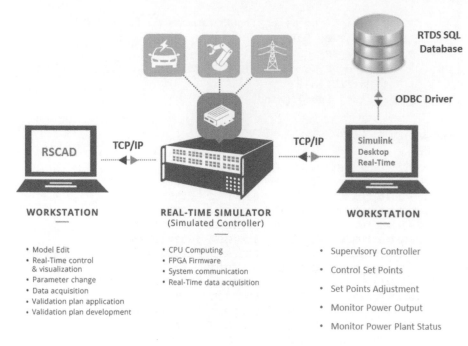

FIGURE 10.35 Hardware-in-the-loop component interaction with real-time SQL database.

```
1     SELECT Battery.DT,
2         Battery.MatlabTime,
3         Battery.B_SetPoint_Norm,
4         PV.PV_SP_Norm,
5         Wind.Wind_SP_Norm
6     FROM ( ( Rtds.dbo.Battery
7     INNER JOIN Rtds.dbo.PV
8     ON ( Battery.dt = PV.dt
9         AND Battery.matlabtime = PV.matlabtime) )
10    INNER JOIN Rtds.dbo.Wind
11    ON ( Battery.dt = Wind.dt
12        AND Battery.matlabtime = Wind.matlabtime) )
13    --WHERE Battery.MatlabTime >= 0
14    --WHERE Battery.MatlabTime >= 0.1667 AND  Battery.MatlabTime <= 0.3333
15    WHERE Battery.MatlabTime between 0.1666 and 0.35
```

FIGURE 10.36 Example of SQL query used.

10.6.5 Developing Matlab Function to Handle Data Fetching

A MATLAB function has been developed to manage the ODBC connection, the SQL query manipulation and execution, and data fetching and structuring. The function reads the simulation time, changes the query accordingly, and fetches the set point data. Later, the fetched data are stored into a local MATLAB table while discarding duplicated entries. The function can be called by setting a periodic trigger or using a function-call trigger within Simulink. To be compatible with the SLDRT toolbox, this function had to be transformed to MATLAB Level-2 S-Function, because this toolbox can only run function written in C/C++ language as well as handling different Simulink simulation events and phases. The following code illustrates this:

```
%Set query to execute on the database
query = ['SELECT Battery.DT, ' ...
    '       Battery.MatlabTime, ' ...
    '       Battery.B_SetPoint_Norm, ' ...
    '       PV.PV_SP_Norm, ' ...
    '       Wind.Wind_SP_Norm ' ...
    'FROM ( ( Rtds.dbo.Battery ' ...
    'INNER JOIN Rtds.dbo.PV ' ...
    'ON ( Battery.dt = PV.dt ' ...
    '       AND Battery.matlabtime = PV.matlabtime) ) ' ...
    'INNER JOIN Rtds.dbo.Wind ' ...
    'ON ( Battery.dt = Wind.dt ' ...
    '       AND Battery.matlabtime = Wind.matlabtime) ) ' ...
    'WHERE Battery.MatlabTime between %0.4f and %0.4f ' ...
    ''];
query = sprintf(query,L); %assign query range from input
sp = fetch(Conn,query);
%construct set points structure
setpoints.wind.time=sp.MatlabTime;
setpoints.wind.sp=sp.Wind_SP_Norm;
setpoints.pv.time=sp.MatlabTime;
setpoints.pv.sp=sp.PV_SP_Norm;
setpoints.battery.time=sp.MatlabTime;
setpoints.battery.sp=sp.B_SetPoint_Norm;
MyStruct = setpoints;
block.OutputPort(1).Data = block.Dwork(1).Data;
block.OutputPort(2).Data = MyStruct;
```

SQL Data Fetch
MATLAB file (level-2)
S-Function1

10.6.6 Results of Fetched and Combined SQL Data

The following results illustrate the process of fetching two consequent data patches and combining them together while eliminating the duplicated entries (duplicated entries are highlighted).

```
data1=execute_sql()   % First Query Range [0 , 0.167]
data1 =
   4×4 table
```

MatlabTime	B_SetPoint_Norm	PV_SP_Norm	Wind_SP_Norm
0	-0.10548	0	0.96056
0.016667	-0.25064	0	0.96279
0.15	-0.25064	0	0.95279
0.16667	-0.12532	0	0.94898

```
%%
data2=execute_sql()    % 2nd Query Range [0.15 , 0.35]
data2 =
  6×4 table
    MatlabTime    B_SetPoint_Norm    PV_SP_Norm    Wind_SP_Norm
    _____    _____    _____    _____

        0.15         -0.25064            0            0.96279
     0.16667         -0.12532            0            0.93898
     0.18333                0            0            0.91517
     0.31667                0            0            0.91517
     0.33333          0.11217            0            0.89385
        0.35          0.22434            0            0.87254
```

- The following code combines and eliminates duplicated entries

```
if (~isempty(data1))
    %check if first value of second query exists in first
query and return its Index
    x = find(data1.MatlabTime==data2.MatlabTime(1));
    %combine first data with non duplicated entries from
second data
    data_combined=[data1(1:x-1,:);  data2
] end
data_Combined =
  8×4 table
    MatlabTime    B_SetPoint_Norm    PV_SP_Norm    Wind_SP_Norm
    _____    _____    _____    _____

         0           -0.10548            0            0.96056
  0.016667           -0.25064            0            0.96279
      0.15           -0.25064            0            0.96279
   0.16667           -0.12532            0            0.93898
   0.18333                 0            0            0.91517
   0.31667                 0            0            0.91517
   0.33333            0.11217            0            0.89385
      0.35            0.22434            0            0.87254
```

The previous results verify that the objectives of this section, which are the connections and configuration between EMS and RTDS, have been completed successfully. The developed tools have been integrated with the secondary controller model that was previously built and tested in Simulink Desktop Real-Time. Previously, the source of the hybrid plant resources set points was loaded at and stored one time at the beginning of every simulation and for the entire simulation period. The set points were fixed and couldn't be changed or updated during the simulation.

Currently, the set points source has been redirected to the output of the newly developed MATLAB Level-2 S-Function, which is called "SQL data fetch." This function executes a single dynamic SQL query to fetch data for the current simulation time from an SQL database. The function is called every 5 minutes and fetches data for the next 10 to 15 minutes. The set points can be updated inside the SQL database independently. Whenever a set point is updated in the database, it will be updated in the simulation model in the next execution of the SQL query. The new fetched data will overwrite any duplicated entries by new updated entries.

REFERENCES

[1] RTDS, "Real-Time Digital Simulator Power System User's Manual," RTDS Technologies Inc, Winnipeg, Manitoba, Canada, 2006.

[2] M. Jamshidi, "System of systems engineering—new challenges for the 21st century," *IEEE Syst. Mag*, vol. 23, no. 5, pp. 4–19, 2002.

[3] E. B. Ssekulima, "Power Dispatch Strategies for Enhanced Grid Integration of Hybrid Wind-PV Power Plants," Masdar Institute of Science and Technology, Abu Dhabi, UAE, 2016.

[4] M. Luo, "Multi-Physical Domain Modeling of a DFIG Wind Turbine System using PLECS ®," *Plexim GmbH, Appl. Ex.*, no. 2–14, pp. 1–13, 2014.

[5] E. B. Ssekulima, "Power Dispatch Strategies for Enhanced Grid Integration of Hybrid Wind-PV Power Plants," Masdar Institute of Science and Technology, Abu Dhabi, UAE, 2016.

[6] S. Müller, M. Deicke, and R. W. De Doncker, "Doubly fed induction generator systems for wind turbines," *IEEE Ind. Appl. Mag.*, vol. 8, no. 3, pp. 26–33, 2002, doi: 10.1109/2943.999610.

[7] J. Fletcher and J. Yang, "Introduction to the Doubly-Fed Induction Generator for Wind Power Applications," *Paths to Sustain. Energy*, 2010, doi: 10.5772/12889.

[8] C. A. V. Guerrero et al., "A new software-in-the-loop strategy for real-time testing of a coordinated Volt/Var Control," *IEEE PES PowerAfrica Conf. PowerAfrica 2016*, pp. 6–10, 2016, doi: 10.1109/PowerAfrica.2016.7556559.

[9] "Software-in-the-loop | SIL testing | SIL software." https://www.opal-rt.com/software-in-the-loop/ (accessed Nov. 13, 2020).

[10] H. T. Nguyen, G. Yang, A. H. Nielsen, and P. H. Jensen, "Hardware- and Software-in-the-Loop Simulation for Parameterizing the Model and Control of Synchronous Condensers," *IEEE Trans. Sustain. Energy*, vol. 10, no. 3, pp. 1593–1602, 2019, doi: 10.1109/TSTE.2019.2913471.

[11] A. Valibeygi et al., "Microgrid control using remote controller hardware-in-the-loop over the Internet," *2018 IEEE Power Energy Soc. Innov. Smart Grid Technol. Conf. ISGT 2018*, pp. 1–5, 2018, doi: 10.1109/ISGT.2018.8403345.

[12] M. Bakhtvar et al., "Dispatchable Renewable Generation : Oman's Harweel Wind Farm Potential," *Cigre GCC*, 2019.

Index

Note: Locators in *italics* represent figures and **bold** indicate tables in the text.